About Island Press

Island Press, a nonprofit organization, publishes, markets, and distributes the most advanced thinking on the conservation of our natural resources—books about soil, land, water, forests, wildlife, and hazardous and toxic wastes. These books are practical tools used by public officials, business and industry leaders, natural resource managers, and concerned citizens working to solve both local and global resource problems.

Founded in 1978, Island Press reorganized in 1984 to meet the increasing demand for substantive books on all resource-related issues. Island Press publishes and distributes under its own imprint and offers these services to other nonprofit organizations.

Support for Island Press is provided by Apple Computers, Inc., Mary Reynolds Babcock Foundation, Geraldine R. Dodge Foundation, The Charles Engelhard Foundation, The Ford Foundation, Glen Eagles Foundation, The George Gund Foundation, The William and Flora Hewlett Foundation, The Joyce Foundation, The John D. and Catherine T. MacArthur Foundation, The Andrew W. Mellon Foundation, The Joyce Mertz-Gilmore Foundation, The New-Land Foundation, The J. N. Pew, Jr., Charitable Trust, Alida Rockefeller, The Rockefeller Brothers Fund, The Florence and John Schumann Foundation, The Tides Foundation, and individual donors.

IN PRAISE
OF NATURE

IN PRAISE
OF NATURE

EDITED AND WITH ESSAYS BY
STEPHANIE MILLS
Assisted by Jeanne Carstensen

FOREWORD BY TOM BROKAW

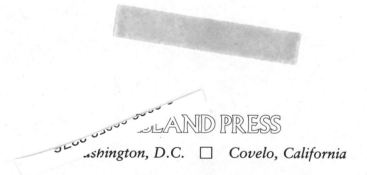

ISLAND PRESS

...shington, D.C. ☐ Covelo, California

The author gratefully acknowledges permission to reprint from *Tao Te Ching* by
Lao Tsu, translated by Gia-fu Feng and Jane English. Copyright © 1972 by Gia-
fu Feng and Jane English. Reprinted by permission of Alfred A. Knopf Inc.

Library of Congress Cataloging-in-Publication Data

In praise of nature / edited and with essays by Stephanie Mills :
foreword by Tom Brokaw : assistant editor, Jeanne Carstensen.
p. cm.
Includes bibliographical references.
ISBN 1-55963-035-3.—ISBN 1-55963-034-5 (pbk.)
1. Environmental protection. 2. Ecology. I. Mills, Stephanie.
II. Brokaw, Tom.
TD170.3.I5 1990
363.7—dc20 90-33875
CIP

Printed on recycled, acid-free paper

Manufactured in the United States of America

10 9 8 7 6 5 4 3 2 1

To the memory of Bob Carroll,
sui generis performance artist whose
prodigious creativity and fertile intelligence
truly bespoke the Earth

Contents

FIRE 89

WATER 121

SPIRIT 155

Publisher's Preface

Public concern about the environment has never been as widespread as it is at this moment in 1990. Around the globe, while schoolchildren plant trees and mothers protest toxic dumps in their neighborhoods, politicians support measures to protect the quality of the air and water. Ordinary citizens tell pollsters that they are willing to pay more for environmentally sound products, and college students enroll in courses about ecology and environmental science, preparing for new careers. If the nineties are truly to become the decade during which we begin to repair our relationship to the earth, there is plenty of evidence that we are off to an appropriate beginning.

In response to this growing public concern—and in an effort to provide a resource for everyone who has a desire to know more about the earth and our relationship to it—Island Press is proud to publish *In Praise of Nature*, edited by Stephanie Mills. The many voices in this book address the deeper questions that underlie the environmental problems that we read and hear about every day in the news. From many different perspectives, this book contemplates the nature of our relationship to the earth. It asks what kind of relationship—what perceptions, what behavior—lies at the heart of an ecologically sustainable life.

Since 1978, Island Press has published books on the environment, with an emphasis on the practical tools and solutions that are needed by public officials, environmentalists, and business and community leaders who are actively at work trying to solve environmental problems. In the last few years, we have noticed two significant developments: first, the demand for environmental information has increased dramatically; and, second, the definition of the kind of information that people need has expanded considerably. Just as concern for the environment now extends well beyond the professional, academic, and activist communities, so also the kind of information needed now includes not only the latest scientific, economic, and practical knowledge, but also insights and commentary about the philosophical, ethical, and spiritual context of environmental issues. *In Praise of Nature* aspires to meet some of this expanded need. It recognizes that until we—as individuals and as a society—know the answers to the fundamental questions and learn to live by that knowledge, legislative and technical solutions to environmental problems will be only temporary. The book opens with a moving foreword by Tom Brokaw, which describes his personal connection to the environment and sets the context for the book. It is followed by an interview with Denis Hayes, organizer of both the original Earth Day and Earth Day 1990. Denis provides us with a perspective on the last twenty years.

The heart of *In Praise of Nature* consists of five sections divided according to the time-honored pattern of the elements of life: Earth, Air, Fire, Water, and Spirit. Each section begins with an essay that touches on the gifts that the element provides the living Earth, the threats to the systems within that element, and the hopes for dealing with those threats. Each essay is followed by reviews and excerpts of ten or twelve illuminating books—ranging from science to poetry to how-to—that address the wide range of topics suggested by each element. A sixth and final section presents brief descriptions of more than one hundred additional books, a sampling of the further wisdom and knowledge available to everyone who wishes to pursue the thoughts offered here.

Stephanie Mills brought an eloquent voice, incisive intellect, and a passionate vision to the conceptualization and organization of the book and its central essays. With the indispensable help of Jeanne Carstensen, she identified books and reviewers representing not only good literature but also diverse and provocative points of view. As is the way with most books, this one grew well beyond its planned time frame. We are grateful to Stephanie and Jeanne for their dedication and commitment to bringing *In Praise of Nature* to life.

It was inspiring and a bit uncanny to read the book reviews that came in from every region of the country, from reviewers who included longtime environmentalists, scientists, legislators, writers, and others. Reading the contributions as they arrived, we were struck by the echoes among them; though each voice is unique, the same concerns, the same themes, the same longings—even, sometimes, the same phrases—appear in many of the reviews. If this sampling of opinion is a harbinger of a larger shift of public understanding, it is easy to be optimistic that we will find our way through the environmental dilemmas that confront us today.

If, as we hope, you are inspired to take some action in your own life or in your community after reading this book, you can find other books that offer guidance on those next steps. You may want to write for the *Annual Environmental Sourcebook* from Island Press, which lists more than 150 books on environmental issues; information about ordering can be found at the back of this book.

—CHARLES C. SAVITT
President, Island Press

Foreword

One of the privileges of my South Dakota upbringing was an early exposure to land, sky, and water unfettered by pavement, development, or other forms of population pressure. As a boy, I lived on prairie bluffs hard by the Missouri River in an area where not so many years before great herds of buffalo roamed and the mighty Sioux nation prevailed. It was a kind of nineteenth-century place in the middle of the twentieth century.

Oh, it wasn't perfect where I grew up. They dammed the Missouri. Small towns had primitive sewage treatment at best. Pesticides and fertilizers were introduced to agriculture without much thought of their consequences. Still, there was so much land, sky, and water the abuses were relatively small—or so they seemed at the time.

Those surroundings had a profound effect on my life. Well before I had heard of Shelley, his song to nature rang through my subconscious.

> *I love snow, and all the forms*
> *of the radiant frost;*
> *I love waves, and winds, and storms,*
> *Everything almost*
> *Which is nature's, and may be*
> *untainted by man's misery.*

Yet I also longed for the big city and bright lights, for the social intercourse that comes with the urban environment more readily than with the rural setting. So I set off for distant skylines.

It was all that I had hoped: intellectual and political energy, a rich diet of entertainment, a place crowded with new faces and new personal and professional challenges. I left behind my environmental roots—for a while.

Then, when the sense of discovery began to fade in my adopted habitat, I developed another consciousness: an awareness of the terrible price of this relentless drive to alter nature for the comfort and convenience of the rapidly expanding population gravitating to urban areas.

Some of the alteration was deliberate and provocative—deserts uprooted and valleys plowed under, rivers dammed, wetlands drained, hillsides paved over. Other forms of alteration were by-products—foul air from too many cars moving too many people over too much pavement; water, the womb of life, contaminated in the name of progress.

It began to close in on me. I returned to my childhood pleasures. I went to the mountains, where I discovered what were for me new cathedrals for my spirit. I made a pilgrimage back to the prairie and stood in awe on its rim, absorbed in a late autumn storm at night, accompanied only by a steady wind blowing through the high grass. I moved to the edge of the sea and felt diminished by its unrelenting strength and infinite forms.

In a way I was reborn.

Since then I have looked at this planet only as a whole place. Most of my professional waking hours are spent documenting political, social, and economic change. The imperatives of daily journalism, alas, allow little time for context and reflection.

Nonetheless, I endeavor daily to measure that political, social, and economic change against the consequences for the environment. Recently we have been privileged to witness a dramatic and historic manifestation of mankind's inherent desire for freedom. I can think of no more appropriate way to prepare for the millennium.

That personal freedom, however, will be greatly devalued if mankind becomes imprisoned by a hostile environment. The assault on the environment is also an assault on our freedom. If nature is corrupted to reward the avarice of only a few, that is a punishment for the masses.

Moreover, there is a moral obligation to protect nature just as surely as there is a moral obligation to promote justice and political freedom. Future historians will measure us not just by our social compacts but by our stewardship of nature as well.

Seneca, unaware of the tools available to modern man, wrote, "It is difficult to change nature." Nature does remind us of her place with earthquakes, fires, hurricanes, tornadoes, droughts, and floods. Still, we have managed to change her, often without much thought, and never for the better.

In these essays you will find the essence of nature as best as it can be captured on the printed page. They will lift your vision and fill your soul with awe. They will remind you that we are here to be nurtured by the environment, not to consume it. We reflect in its glory. It should not be shamed by our indifference and ignorance.

—TOM BROKAW

Grassroots Momentum Toward a Sustainable Future

AN INTERVIEW WITH DENIS HAYES

Denis Hayes is a Harvard Law School dropout whose outside agitating in the early seventies was national in scope and consequence. Nefarious? Hardly. It was Hayes who organized the first Earth Day, in 1970. Rising to a similar occasion in 1989, Hayes took a leave from a California law firm (having finished his degree at Stanford University) to serve as chief executive officer of Earth Day 1990, a twenty years' more sophisticated (and accordingly more expensive) organizing campaign to launch an environmental decade.

Modest, thoughtful, and soft-spoken, Hayes has an uncommon depth and diversity of experience in the environmental mainstream. Between Earth Days, he founded Environmental Action, lobbied for the Clean Air Act, was a senior fellow at the Worldwatch Institute, wrote *Rays of Hope: The Transition to a Post-Petroleum World*, chaired Sun Day, was the director of the federal Solar Energy Research Institute, and has been a visiting professor at Stanford University.

In conversation at a cafe near the Earth Day 1990 office in Palo Alto, California; on paper; and over the telephone, I queried Hayes for his thoughts about Earth Day and the shape of environmentalism itself.

Asked what the first Earth Day accomplished, he responded, "A whole raft of legislation was passed: the Clean Air Act, the Clean Water Act, the Environmental Education Act.... The Supersonic Transport was stopped. The use of carcinogenic herbicides in Southeast Asia was halted. The federal Environmental Protection Agency was created. Seven of a 'dirty dozen' congressmen were defeated in the fall of 1970. Most national environmental groups doubled or tripled in membership.

"Today," he continued, "the big new issues—aside from toxic and hazardous waste—tend to be international in scope. Global warming. Ozone depletion. Rainforest destruction. Acid rain.

"These are tough issues, in that their solutions tend not to be congruent with political boundaries. One country gets the benefits; a different country bears the costs. If the ocean rises a few feet because of global warming, the Maldives Islands will be submerged. The islands' contribution to global warming is infinitesimal, but they will pay a huge price—as will the rice-producing river deltas of eastern Asia.

"The deterioration of the global commons is more difficult to address than are local problems. The United Nations Environment Program is modestly funded and has no enforcement powers. For all the flaws in the U.S. Environmental Protection Agency, it has the necessary muscle to solve many national problems if only it had the will. The world desperately needs an international agency with real power to address the transboundary environmental problems."

In subsequent telephone conversations, Hayes discussed the reasons he feels the 1970 success can be replicated on a global stage in 1990. "The parallels between the late 1960s and the late 1980s are really striking," he maintains. "Then we had the Santa Barbara oil spill; now we have the *Exxon Valdez* spill. Then we had Rachel Carson calling attention to pesticides; now Meryl Streep is carrying the same message to a broader audience. Then we had ugly smokestacks and urban smog on the evening news; now the evening news is filled with toxic wastes and global warming. Then and now, people are profoundly worried about the future of the planet.

"A less obvious parallel is also worth noting. Both then and now the world experienced a rising tide of activism. The late 1960s saw the French and Mexican student uprisings, the Chinese Cultural Revolution, and the antiwar movement. Today we have the political liberation of Eastern Europe, the rapid worldwide growth of Green political forces, and the women's movement. And the forces of reaction are always with us. Then it was the brutal Soviet invasion of Czechoslovakia; now it is Tiananmen Square.

"The most interesting political strands currently in the wind are resurgent democracy, the winding down of the superpower arms race, and a new concern throughout the world for sustainable societies. A goal of the modern environmental movement should be to weave these strands together around an environmental ethic that could become the most compelling ideological force in the twenty-first century."

Hayes believes that this new environmental ethic is evolving more rapidly in practice than in theory. "The masterwork that would apply ecological principles to just, peaceful, sustainable human societies has not yet been written, but a lot of practical progress in that direction is being made nevertheless.

"The most exciting environmental progress, in the United States and abroad, is being achieved by diverse, decentralized, grassroots organizations" according to Hayes. "The forces now gathering momentum in the United States have the same roots as those recently unleashed in Eastern Europe, the Philippines, China, and elsewhere. People are fed up with governments that lie to them, and with governments that don't respond to their most basic needs."

Hayes argues that the most responsive governments are those that are closest to the people. "The most innovative programs are those at the neighborhood, city, and county levels. Bans on CFC emissions are being enacted; toxic incinerators are being defeated; six-lane freeways are being stopped in their tracks; and curbside recycling programs are spreading like kudzu.

"These environmental victories are the result of the decentralized democracy that Jefferson championed so eloquently. It is the sort of democracy in which people meet with their elected leaders instead of relying on fifteen-second image ads for insight. It is a democracy in which campaign promises mean something.

"Today, American national politics is dominated by forces that value 'stability' above all other values. Our international policies too often encourage that stability by buttressing Marcos, Deng, the Shah, and their ilk against the democratic impulse.

"Stability makes no sense in a world that is on a manifestly unsustainable course—a course destined to lead to ecological collapse unless dramatic changes are made. And our support for 'stable' despots is not only offensive to American traditions, but also self-defeating as a geopolitical strategy. If enlightened despotism ever had a place in the political firmament, that time was long ago. Today, despotism means Idi Amin—not Marcus Aurelius.

"The world is experiencing a resurgence of democracy unparalleled since the eighteenth century. In our dull pursuit of stability, we repeatedly lend our support to history's losers. Although this 'hard-nosed geopolitics' is always rationalized by the Kissinger types as serving our national interests, it is a betrayal of our national purpose."

Leaving the global theater and returning to practical politics, Hayes lamented the failure of American environmental organizations to find more common ground. "Often grassroots organizations dislike and distrust those operating at the national level. Similarly, Washington-based groups often view local organizing as little more than busywork. There are other splits as well.

"I understand the differences between the deep ecologists and the social ecologists, between the monkeywrenchers and the feminists, between the Greens and the environmental PACs, between the NIMBY groups and the Washington groups. But they have more in common than they admit, and they squander scarce resources by constantly shooting their biggest guns at one another.

"The real hope in all this may rest, as it did in 1970, with the youth. They are not yet into ego cults or ideological camps. They simply want to have a future. Students have led democratic uprisings around the world, and they seem to be playing a widening role in the United States. The most open-minded but wildly enthusiastic crowds that I address these days are on college campuses. Maybe this next generation of activists will avoid some of the pitfalls that plagued my generation."

In particular, Hayes expressed the hope that, unlike in 1970, the environmental movement this time would embrace a wider agenda. "Our principal achievements during the last twenty years have been political. That's fine as far as it goes, but we need to paint with a broader brush. We need to involve people as consumers, investors, workers, and parents. A sustainable society cannot be created by lobbyists and lawyers. It will require a cultural revolution that transforms the economy, reshapes lifestyles, alters the educational system, and offers people new avenues of social and spiritual fulfillment.

"The budding international environmental movement must hitch its wagon to a far-reaching set of ideas—to an inspiring vision—if it is to realize its potential to transform the world. It must be vibrant, creative, and playful, while at the same time holding its central values inviolable.

"Perhaps most important, this time we must not make the mistake of asking too little of ourselves."

—STEPHANIE MILLS

Prologue

Earth Day 1990 was a Janus-faced event: It harked back to a moment of environmental awareness two decades ago and sought to inaugurate a decade, at least, of social change to save the planet.

So the occasion for this book was the endeavor of two generations of Earth-activists, and the content of this book spans the wisdom of even more. Inspired by the *Whole Earth Catalog*'s respect for individual intelligence and initiative, properly equipped, we too are providing access to tools. Durable goods, the books reviewed herein aren't, with few exceptions, the latest. Rather they are the best statements of certain big ideas that must be mastered in order to make sense of the situation we confront, and having confronted it, to change it for the better. Ecological thinking has a long history, and it would be a waste and a shame to allow its nobler works and authors to be stampeded past.

In addition to reflection and change, renewed environmental awareness calls for a celebration of the elements of life on this planet: Earth, Air, Fire, Water, and Spirit.

> ... the seemingly passive but actually living soil, which tendered correctly, can foster the life contained in a tiny seed and husband the rain
> ... the trembling of leaves, rippling of grasses, the tracks left by invisible winds sweeping the Earth in great gyres, ushering the weather
> ... the fascination of flame leaping gold and blue from chunks of wood as they blacken, vermillion to sere ash, warming cold wet bodies with their sacrifice

. . . the color that is no color and sounds that are sublime of living water coursing downhill or jetting up from within the Earth to be tasted pure from a dipper at a spring

. . . the marvels of spirit, of loving acts, and giving; of mutuality and imagination—not just the traits of the human species, but of sperm whales buoying up a wounded sister, of bees fanning the hive.

These are marvels that need not transcend the Earth: They have their existence within it completely. May they always continue.

A book is only a book. What you have in your hand is just an artifact of the effort to make words on paper serve life on Earth. This book is an invocation of the powers of the Earth and a chronicle of the eloquence and intelligence they have long been inspiring. It is to seed change within and without.

EARTH

I T'S ALL TOO EASY, in the dailiness of existence, to begin to take life for granted. And yet that there is life, and that life comes together in so many forms, from so many elements and smaller forms, is never less than astounding.

"A mouse is miracle enough to stagger sextillions of infidels," wrote Walt Whitman. Life on Earth is the magic you can watch minute by minute. Every single cell is a wonder, a cooperative community of subcellular bodies that have come together in the destiny of being a larger metabolic unit. How many millions of cells, how different one to the next, differentiated from one tissue to the next, how many of those miracles compose the unity that is a trembling field mouse? In one human body, we know, there are a hundred times more cells than there are stars in the galaxy.

For oxygen to be present in Earth's atmosphere in an amount useful to air-breathing life-forms took eons. The development of soil and its fostering of millions of species required billions of years. Vascular plants, among them the grasses that make all flesh, are a sophisticated development. Woody stems, root hairs, capillary action drawing minerals and water up toward leaves for transformation through photosynthesis—flowers and seeds are inventive measures for life to have taken in its ongoing play. We all live by that greenness: Photosynthesis is the limiting factor for all other life on Earth. Even though we may procure it at a supermarket, our sustenance is entirely from nature, and in nature it cycles through. Eat or be eaten is the name of the game. There's no sustenance without something's dying, the dying in parts of perennial plants, the dying

3

of prey in the jaws of a predator, the dying of predators and their reduction by the actions of myriad soil makers. Death drives reproduction, the paramount force in evolution. Transmission of genetic information is the aim of each being, transmission of characteristics and admitting to the possibility of change—gills to lungs, limbs to fins, the darkening of a moth species' protective coloration to adapt to a rain of industrial soot.

We don't know how many different kinds of life-forms there are. Estimates of the numbers of species range from five to thirty million. Because so many are as yet unclassified, microscopic, or obscure, nobody is quite sure how many there are actually. However many there are, habitat destruction—the clearing of tropical forests, the logging of the ancient forests of the Pacific Northwest, the dredging and filling of wetlands, the smothering of coral reefs with effluents from development onshore—is quickening the pace of extinction in our time. Epochs of extinction have occurred before in Earth's history. The late Permian extinction removed half of the families of marine life, and the greatest mass extinction to date took place in the late Cretaceous period and wiped out the evolutionary lineages of the dinosaurs. The human species is on the verge of causing a mass extinction to top that, an evolutionary event the like of which hasn't been felt on Earth in sixty-five million years, taking an immense toll in genetic diversity. It's a dying on an inordinate scale, but we can take action to stop it, once we get, in our hearts and our bones, that our fate as a species is bound up with that of every other creature: We live one life.

Our early ancestors understood themselves to be as much a part of the Earth as anything else in their world, and felt that everything was alive and invested with power and mystery. A sense of the Earth as a great goddess, a bringer-forth of life and an enfolder of the dead, seems to have been widespread in Paleolithic and Neolithic times. Creation myths envision the Earth as a being—sometimes as a great animal. Native Americans saw their part of the Earth to be Turtle Island. They had a sense of the Earth itself as being vital and long-living, as turtles are, tenaciously enduring. A turtle's heart can beat for a long, long time.

Until well into the Middle Ages, even most Europeans believed that Earth was, in a sense, alive. The Enlightenment, with its clockwork sense of matter, dismissed that understanding from civilization until the late twentieth century, when scientists James Lovelock and Lynn Margulis announced their Gaia hypothesis. Oversimply stated, the Gaia hypothesis is that together the planet, its life-forms, and its atmosphere are interacting and mutually creating, and have some of the properties of living tissue; that the Earth is like an organism. Thus, the Earth is not just a molten-centered ball of rock on whose surface life fortuitously happened to evolve. Rather Gaia (an archaic name for the Earth mother) lives.

Having access to great bodies of relatively new and rapidly developing scientific understanding is a sure blessing of our time. Geology, for instance, which emerged in the late eighteenth century, challenged the religious dogma of creationism by asserting that past geological changes were brought about by the same causes presently working on the Earth's surface, and by stressing the near-infinite slowness of the process of Earth-shaping change. Physical evidence that the Earth's surface was being unmade and remade, that continents were worn down and mountain ranges thrust up over the eons of time, was contrary to scripture, and blasphemous.

The imagination it takes to grasp the spans of time it took for the planet's surface to cool down, and to slow down, and for there to begin to be a less cataclysmic shaping of the surface than by meteor bombardment; and to picture the dance of the plates, bearing along the continents, and, with them, the arrays of species that would diverge and change once they were separated by new oceans and landforms, as the forests of Asia and the forests of North America did, is wilder than the imagination it takes to believe the myth that it was all made in six days.

Earth seems ancient; yet there was a time when Earth did not exist, and then a long time, 750 million years, give or take a few, when it was inhospitable to life; then billions of years when Earth was amenable only to microbial life, whose presence, over these great spans of time, transformed Earth's atmosphere and made it possible for yet more complex forms of life, like nucleated cells and multicelled plants and animals, to evolve.

We now believe that Earth was formed some 4,600 million years ago, from the collisions and mergers of small planetary bodies that were spun off as the sun condensed. The formation of the Earth's crust and its oceans took about 800 million years. Over the first 4 billion years, the Precambrian era, Earth's permanent crust was formed, deposits of metals and massive formations of metamorphic rock laid down.

Then weathering and erosion began wearing down the mountain ranges with wind and rain; sedimentation began laying down sea bottoms and continental shelves.

Next came the Paleozoic era, the age of invertebrates and marine life. At its beginning, in the Cambrian period, which commenced 600 million years ago, the continents were covered by shallow seas whose floors would become sandstone, shale, and limestone. Then for the 65 million years of the Ordovician, North America, Europe, and Africa moved together. About 350 million years ago, in the Devonian period, Europe and North America collided once again, churning up first the Appalachian, then the Ouachita and Marathon mountain chains. During the muggy Carboniferous period 345 to 280 million years ago, forests that would eventually be transformed into coal sediments were flourishing in great swamps; more continental collisions were taking place.

Geologic time marched on, through the Mesozoic era, which lasted from 230 to 65 million years ago. The climate changed; Pangea, a supercontinent, formed and began to break apart again. The ocean basins opened; the Appalachians were worn down, and the Pacific plate cruised under the North American plate (plates are the vast puzzle pieces, floating on Earth's core of molten rock, that, according to tectonic theory, carry the continents and portions of the ocean bed slowly and inexorably about the planet) and kicked up the mountains of the West; and then the continents were flooded. Dinosaurs roamed.

Inching closer to our own moment, the Cenozoic era, age of mammals and seed-bearing plants, began just 65 million years ago. In the beginning, the seas retreated. Volcanism created the Rockies, and weather began to erode them. A mere 23 to 5 million years ago, such timeless features as the Himalayas, Alps, and Andes were being built up; Earth's climate began to cool. During the Pliocene period the polar ice caps were growing, and the familiar outlines of the North American continent appeared. The earliest evidence of humanlike apes and human artifacts dates from the Pliocene period of the Cenozoic. Next came the Pleistocene period—the Ice ages—with snowfalls accumulating over ever-colder summers to become glaciers—rivers of ice grinding rocks to powder, reshaping the surface of continents, and carving deep lake basins during their advances and retreats, all the while pushing around the ranges of plants and animals. This epoch continued from 1.8 million to 10,000 years ago. (Indeed, some theorists argue that our present time is just an interglacial episode.)

Strange to think about the dynamism of the face of the Earth, its continual transformation, and of the profusion and diversity of life taking hold everywhere it can. No two places in nature are identical; every inch of living Earth differs from every other, and in ways that may be fateful for their inhabitants. A particular forest—maple-beech or tropical rain—cannot just spring up anywhere. Ecosystems result from different combinations of soil, slope, elevation, precipitation, and proximity to pole or equator. The living creatures inhabiting ecosystems may be indigenous rarities—fish found only in a few desert hot springs, unable to survive anywhere else, or a salamander wanting to dwell in sluggish peace under a moist rock by an Olympic forest stream whose riffles are kept at just the right velocity by the impediment of occasional fallen giant firs. They may be more common and adaptable—like coyotes and opossums, two species that are currently expanding their ranges. Whichever the case, each species has a unique lifeway and an essential relationship to the health of the whole system, a role to perform.

Humanity is flatly incapable of replicating anything so complex, fortuitous, and subtle as an ecosystem. Ecosystems are the Earth's way of maintaining a dynamic equilibrium among groups of organisms in particular locales: not unchanging, but diverse enough to be able to integrate the changes taking place over evolutionary time. The more diverse the ecosystem, the greater its resilience.

Diversity is a measure of the numbers of species present. Thus species extinctions simplify, and destabilize, whole ecosystems. Because each ecosystem is different, quite place-specific, no formula counsels how far we can push the forgiveness of ecosystems in our desire to exploit the resources they provide, or that underlie them.

As a tool-using, cosmopolitan omnivore, the human animal perhaps finds it difficult to understand how inseparable some creatures are from their ecological niches, how finely attuned to specific temperature ranges, nesting spaces, diet, and territorial needs. Change one too many of these critical variables and a species vanishes. Even before the last member of a species vanishes, its numbers may decline below a level at which there's sufficient genetic variation to allow its evolution to continue. Even though it may be possible to preserve specimens of endangered species in zoos, some qualities, real and ineffable, vanish in captivity. As David Brower commented on the decision to resort to a captive breeding program for condors, "A condor in a zoo may feel a breeze, but it will never know the wind."

The human species has, for the most part unwittingly, committed habitat and species destruction from the beginning. There is some speculation that overhunting by native ancestors may be accountable for the dying-off of the great mammalian fauna—such as mammoths and giant ground sloths—of North America during the Pleistocene era. The Biblical cedars of Lebanon are long gone. The deforestation of the Mediterranean landscape was being deplored even in antiquity. Much habitat destruction has been a result of our (sometimes willful) ignorance of the reality that in changing nature we can never do just one thing.

Science is only now beginning to transcend the boundaries of different disciplines in order accurately to describe relationships occurring in the natural world. Scientists are coming together to mesh their expertise and to grapple with some of the questions raised by ideas like the Gaia Hypothesis, which describes the working of an immense, complex system, and, more urgently, to consider the evolutionary threats posed by the elimination of species-rich habitats like ancient temperate and tropical rain forests and coral reefs.

When we talk about the need for science to develop the knowledge necessary to preserve biodiversity, we must remember that this essential knowledge—what Aldo Leopold termed a land ethic—has been the lifeway of aboriginal cultures. Thus what we are really urging is to have that knowledge expressed, and possibly enhanced and detailed, by science, which seems to be the sacred language of our culture.

Today wilderness and ecosystem preservation has become more than a matter of fighting for legal protection, or acquisition, of a particular habitat (although those alone are no mean feats). Now, if anything like wilderness is to remain, biospheric threats must be confronted as well. Acid rain and greenhouse warming, which by changing the pH of lakes and soils and altering the climate, may compromise whole ecosystems whose members are unable to adapt to such radically changing conditions.

For any attempts to address these problems to have some chance of succeeding, there must be effective control of the growth of the human population. There is an undeniable correlation between the exponential growth in human numbers, the loss of topsoil and trees, and the decline in the population size and overall numbers of virtually all other species.

This correlation isn't exact: Culture and consumption patterns, finally, determine the per capita impact of humans on the biosphere, complicating the population problem with the inarguable need to achieve a more just distribution of the world's goods.

There is little hope for population control without an equal commitment to meeting basic human needs. Many ecologists argue that planning for family sizes below replacement levels (which means a one-child family), and thus an eventual decline in human numbers, is crucial. Granted that the factors that have spurred population growth are complex, and family limitation changes individual lives profoundly, some reduction in family size is generally desired by the world's women. Provided that the commitment to meeting basic needs and fostering self-help development strategies is honored, a decline in overall numbers of children born, in favor of greater quality of life for children surviving, can be positive for human, as well as other, life.

If the individual's stake—and role—in population control is rarely grasped, and acted upon, simply because immediate self-interest usually overpowers any feeling of responsibility for the collective interest, the human stake in ecosystem health has been the informing idea of most preagricultural people. This principle once again is painfully clear to villagers throughout the less-developed countries, and increasingly to national governments. The awareness that the ecology is the economy is beginning to dawn, quickly for subsistence peoples, more slowly for urbanites. Third World villagers hug trees, plant greenbelts, resist being relocated to make way for mining and hydroelectric developments. In the United States the threats to ecosystems and biodiversity are the same, but the perception that these threats threaten us as well is not so keen. Here, ecosystem defense has been given impetus by aesthetic concern.

North America, with its spectacular mountain ranges, coastlines, Great Lakes, and deserts, has a dazzling array of ecosystems: more wilderness to treasure than the Old World knew, and, since the arrival of the Europeans, a fiercer pace of wilderness degradation. All this engendered a strong conservation movement, with visionaries and just plain citizens organizing to defend beloved places. John Muir's losing battle to prevent the drowning of the Hetch Hetchy Valley, which rivaled the Yosemite in its beauty, is one of the turning points of American environmentalism. So, too, the creation of our system of national parks, wildernesses, and national monuments.

U.S. conservation activists fighting mining, road building, clear-cutting, and other inappropriate uses of public lands have resorted to tactics ranging from letter writing to lobbying to litigating to legislating, and as the assaults on remaining wilderness escalate, to non-violent direct action—sometimes risking life and limb.

Activists, first in New Zealand, with the Values Party, and later in West Germany, then Sweden and Italy, have understood environmental destruction in a larger sociopolitical context and started Green political parties whose aim is to create ecologically sustainable societies. Green candidates have been elected to a few seats in national parliaments, where they advocate a whole range of progressive, pacifist policy changes. (In the United States there have been successful Green candidacies at the municipal level, but our two-party system makes the emergence of a successful national Green party unlikely at present.)

It may be that we will be able to look back at the last half of the twentieth century as an extraordinary turning point in Earth's history. These have been decades during which the human species became, and then became aware of being, a geophysical force on the planet. We have, in thirty years, effected atmospheric change nearly as dramatic as that made by the cyanobacteria, which took two billion years to produce enough oxygen for a whole biosphere of life to breathe. Through deforestation and subsequent desertification we have had a geologic impact more widespread than any chain of volcanoes erupting or plates colliding.

We haven't fully comprehended the immensity of this power, and urban culture is not yet quite aware of the absolute necessity to learn the language of evolution—spoken by natural history—if there is to be a positive outcome of human history. For many people, the nearest opportunity to learn a little natural history may be in a park or a vacant lot. And while there is wonder everywhere there is life, it is critical to learn the difference between the hodgepodge of native and exotic species that may be found in a park or vacant lot and the coherence of relatively intact ecosystems. Basic ecological literacy would demand that everyone have some sense of the difference between ancient forests and second- and third-growth woodlands, because life hangs in the balance of knowing those details.

Besides, the experience of learning the details, which means going outdoors, is what can hearten us to segue from the twentieth to the twenty-first century in a lifesome way. As the winter solstice of 1989 approached, snows in great variety fell almost every December day where I live, with deep white quilts laying a silence on the Earth. The soil might not have been frozen then, but the cold suggested a stillness in Earth's pores, where insects and microbes dwell. The pine boughs drooped with the weight of snow, and occasionally a red squirrel's hectic travel from branch to branch would dislodge a powdery clump and send it plopping down with a surprisingly audible impact.

That summertime sense of life's teeming was all but absent now. Cottontail tracks across the drive disclosed their presence, and a mid-December sight of two foxes, ambling and snuffling and scent-marking around the margins of the meadow, was a rare blessing. Conifer needles glimpsed through snow were the only witness to photosynthesis, and mammal footprints the only signs of wildlife. Otherwise, Earth's fertility, diversity, and activity all seemed hidden—and becalmed.

Except for the chickadees and nuthatches. Now these little birds are exceedingly commonplace, and you don't have to be Saint Francis to get close to them, especially if you are willing to top up the bird feeder regularly. Which is what I was doing, all bundled up

11

behind my house one morning, wonderstruck by the whirring fervor of wings around me in all the hush. The longer I stood watching and listening, the more black caps and white cheeks I noticed glinting among the pine branches, perching and waiting politely for the chance to land at the feeder and carry off a black sunflower seed. To sustain their fast darting lives in the cold, they must feed constantly, stoking the tiny metabolic furnace with industry all out of proportion to their size.

Thus in the dead of winter I found myself surrounded by life, and the white loveliness of the December landscape. Hardly a wilderness experience out back, and a monochrome-minimalist antithesis to the immense green riches of biodiversity imperiled in the tropical rain forests. Nonetheless, all that hungry activity was a powerful spate of Earth magic.

The chickadees' chattery insistent voicing of life's chant, it seemed to me, merited no less a response than the oath of the bodhisattva, "sentient beings are numberless—I vow to save them."

Microcosmos: Four Billion Years of Microbial Evolution
Lynn Margulis and Dorion Sagan
1986, Summit Books

So what's new, important, and different in recent ecology? Not the fluctuations of foxes and rabbits. Not the seasonal march of farm fields to old-growth hickory and oak. Not the genetics of fruit flies. The cutting edge is microbes—the invisible life-forms that inhabit everything from our intestines to the deepest vents inside the planet's crust. *Microcosmos* is the history of the "microbial beast" that may just control the composition of the air we (you) breathe, right now. A beast whose beauty may save us from global warming, whose biotechnology invented photosynthesis and fermentation.

Microcosmos is wonderful reading. At times, raucous, catty, joyful, teasing. No other book in biology has this perspective: charting the evolutionary trails of one-celled creatures to today's mind-boggling cell conglomerates. No other book so persuasively switches the old Darwinian "tooth-and-claw" emphasis to the importance of symbiosis, mutual dependence, cooperation, and cohabitation. The vision has widened rapidly in the last decade with the Gaia Hypothesis (life is the inseparable governor of atmospheric stability and change). Lynn Margulis is one of two scientists who first made the leap from microbes to microbial cosmos.

—PETER WARSHALL

"So significant are bacteria and their evolution that the fundamental division in forms of life on earth is not that between plants and animals, as is commonly assumed, but between prokaryotes—organisms composed of cells with no nucleus, that is, bacteria—and eukaryotes—all the other life-forms. In their first two billion years on earth, prokaryotes continuously transformed the earth's surface and atmosphere. They invented all of life's essential, miniaturized chemical systems—achievements that so far humanity has not approached. This ancient high *bio* technology led to the development of fermentation, photosynthesis, oxygen breathing, and the removal of nitrogen gas from the air. It also led to worldwide crises of starvation, pollution, and extinction long before the dawn of larger forms of life."

The Flamingo's Smile: Reflections in Natural History
STEPHEN J. GOULD
1985, W. W. Norton and Company

There's a good deal of the eighteenth century about Stephen Jay Gould. As a scientist, his natural habitat is the museum, in the tradition of Buffon and Cuvier, rather than the field or laboratory. *The Flamingo's Smile* contains one essay about Gould's fieldwork on land snails in the Bahamas ("Opus 100"), and it is presented apologetically, as an anniversary self-indulgence, although it's one of the liveliest pieces in the collection.

Romanticism seems to have missed Gould: He demonstrates neither its emotional identification with wild nature nor its distrust of civilization. Perhaps this is why he is not a conservationist or environmentalist in the manner of fellow scientists Daniel Janzen or Paul Ehrlich. Except for passing references, one would not know from Gould's essays that some of the organisms he writes about so fascinatingly are threatened. They seem suspended in the timeless, untroubled illumination of the display case.

Yet it is precisely Gould's serene conservatism that makes him invaluable to the environmentalist viewpoint. There is no more effective intellectual defense against civilization's exploitative and spendthrift tendencies than a detailed understanding of evolution, and of the history of human thought about evolution. Nobody presents these better than Gould. It is hard to imagine anyone who has read his essays still clinging to the creationist's assumption that the planet is simply a pile of resources put here for human convenience.

—DAVID RAINS WALLACE

"Contrary to the romantic image of science and exploration, many important discoveries are made in museum drawers, not under adverse conditions in the parched Gobi or the freezing Antarctic. And so it must be, for the nineteenth century was the great age of collecting—and leading practitioners shoveled up material by the ton, dumped it in museum drawers, and never looked at it again."

"We live in a world of history and change. As creatures of habit who feel comforted by the discovery of order, we search for principles that grant time a direction—that admit a bit of order into the buzzing and blooming confusion of history. But arrows of time are hard to find and science hasn't given us many. The second law of thermodynamics, with its increasing entropy and decreasing order in closed systems, is our most famous agent of direction. Most proposals from evolutionary biology are spurious and based more on our hopes and expectations than the workings of natural selection—the notion of continual progress in particular."

"My profession embodies one theme even more inclusive than evolution—the nature and meaning of history. History employs evolution to structure biological events in time. History subverts the stereotype of science as a precise, heartless enterprise that strips the uniqueness from any complexity and reduces everything to timeless, repeatable, controlled experiments in a laboratory. Historical sciences are different, not lesser. Their methods are comparative, not always experimental; they explain, but do not usually try to predict; they recognize the irreducible quirkiness that history entails, and acknowledge the limited power of present circumstances to impose or elicit optimal solutions; the queen among their disciplines is taxonomy, the Cinderella of the sciences."

CARL ORTWIN SAUER

Carl Ortwin Sauer was born of German immigrant stock in Warrenton, Missouri, on Christmas Eve, 1889. Sauer completed his schooling in the local college and ultimately moved on to the University of Chicago for graduate school, attracted by the concern there for the natural processes and regional studies. After a year or so he was appointed to the Department of Geography at the University of Michigan and rose to chairman in seven years. In 1923 he was called to the University of California as chairman of the Geography Department and remained at Berkeley for the rest of his career. Nevertheless, he always considered himself a son of the Middle Border, and ultimately he was returned to the family plot in the Warrenton cemetery.

In 1943 Sauer wrote in a letter, "The treasure a scholar lays up on earth is largely the printed page." For most of his professional life Sauer wrote essays, prefaces, and reviews and did not start writing books until he had retired. His shorter works have been gathered in two books: *Land and Life,* edited by John Leighly (1963; 1965, University of California Press), and *Selected Essays 1963–1975* (1981, Turtle Island Foundation), edited by Bob Callahan. The latter collection ends with "The Agency of Man on the Earth," Sauer's introduction for the great Green book *Man's Role in Changing the*

16

Face of the Earth, edited by William L. Thomas (1956, University of Chicago Press), based on the Wenner-Gren Symposium of 1955. Sauer was the principal organizer of this symposium, and this book is one of the foundation stones for the present environmental movement.

Sauer also wrote "a first book in geography," *Man in Nature: America Before the Days of the White Men.* This book, originally published in 1939, was the first of a planned series for elementary school texts. Professor Sauer was requested by the publisher to undertake this task, with the assistance of a professional educator. Now, fifty years later, it is not known how much of the original language was modified in this process, but the spirit and approach of this book are Sauer's, a landmark treatment of his concept of geography as human ecology.

In 1975 a facsimile paperbound copy was produced by the Turtle Island Foundation. In 1980 a "second edition" was produced, with a clear statement of the publication history and a two-page introduction by Bob Callahan, mostly about Carl Sauer. The book has now become very difficult to obtain.

Sauer's books were written out of love for his subject; ostensibly historical, they always expressed interest in the ecological aspects of man's presence on the Earth: *The Early Spanish Main* (1966, University of California Press), *Northern Mists* (1968, University of California Press), *Sixteenth Century in North America* (1971, University of California Press), and *Seventeenth Century North America* (1980, Turtle Island Foundation). All of these books were published after he was seventy-five years old! Sauer left us quite a legacy, and his style of history has had its influence. (The most recent and important example of this approach is William Cronon's *Changes in the Land: Indians, Colonists, and the Ecology of New England* [1983, Hill and Wang], which concludes with a quote from Carl Sauer: Americans had "not learned the difference between yield and loot.")

Sauer also played an important part in private foundations and government committees. He wrote the major part of the 1934 report of the President's Science Advisory Board that resulted in the establishment of the Soil Conservation Service. Sauer was above all a teacher, who knew what a study of the land should be.

17

In this observation of the twentieth anniversary of Earth Day, we should note that this is also the centenary of Carl Sauer's arrival among us, for he is among the guiding geniuses of our environmental conscience.

—JOEL W. HEDGPETH

Land and Life
1963; 1965, University of California Press

The Early Spanish Main
1966, University of California Press

Man in Nature: America Before the Days of the White Men
1939; 1975, Turtle Island Foundation

"The rise and fall of cultures or civilizations, which have interested most historically minded students of man, cannot fail to engage the historical geographer. A part of the answer is found in the relation of the capacity of the culture and the quality of the habitat. The case is relatively simple if destructive exploitation can be shown to have become serious. There is also the knotty problem of overpopulation (which may be very much a reality in the culture-historical sense though a heresy to the theoretical social scientist), involving diminishing opportunity and sharing for the individual. There may arise loss of productive energy by maldistribution of population as between country and town, between primary producers and those who are carried as leisure class. There may be a shift of comparative advantage to another people and area. A melancholy and stimulating subject is this scrutiny of the limits of culture."

—Land and Life

"Dogs that did not bark were noted by Columbus. These were found throughout the islands, and later around the Caribbean and beyond. They were small, Las Casas said, of the size of a lap dog, and were eaten,—indeed, were kept to be eaten. Oviedo thought they

18

were used in hunting, an unconfirmed statement of marginal probability. In the starving time of 1494 the Spaniards ate them up, it was said, which could have been true only of the northern district at the time. At any rate, they became extinct on the islands before the Indians did, an early loss in a culture yielding to pressure. At a guess, this dog was both family pet and entrée served up at festivals, as it was in Mexico."

—The Early Spanish Main

"We think it is a good thing to know about Indian days. We could not live like the Indians, even if we wished to do so. We have our own way of living. But we did not need to cut down so many forests, and we did not need to destroy so much wild game. Often we have made the land poor and ugly. The land was natural and beautiful in Indian days. Perhaps we should make parts of it look once again as it did in Indian days."

—Man in Nature

Arctic Dreams: Imagination and Desire in a Northern Landscape
BARRY LOPEZ
1987, Bantam Books

When I am asked to recommend one book on the Far North, my immediate and unwavering choice is *Arctic Dreams*. For several years, author Barry Lopez wholly immersed himself in the Arctic, traveling by skin boat and tanker ship, dog team and pickup truck, airplane and footpath; in the company of Eskimo hunters and biologists, bush pilots and ships' captains, oil workers and environmentalists. The result is a physical and mental journey—prodigiously researched, eloquently written—a celebration of nature and humanity conjoined at the poleward extreme.

19

In *Arctic Dreams,* Lopez demonstrates that hard science, precise history, and serious ethnography can be written in a style that is both accurate and elegant. His book portrays the splendor of an immense, hard-bitten, wild terrain, the beauty and tenacity of life on the tundra and perennially frozen sea, the cleverness and wisdom of the Eskimo people, the intrepidity and blundering of European explorers, and the paradoxical mix of love, insensitivity, fascination, and greed that has carried modern civilization into the hyperborean zones.

But this book is more—much more—than a regional natural history. In his personal musings and his descriptions of Eskimo culture, Lopez shows us a way that humans can use the environment and live as responsible members of a natural community, by mediating their behavior through canons of reverence and restraint. As he evokes this ancient and earthbound wisdom, he urges us, tenderly, to listen.

—RICHARD NELSON

"I took to bowing on these evening walks. I would bow slightly with my hands in my pockets, toward the birds and the evidence of life in their nests—because of the fecundity, unexpected in this remote region, and because of the serene arctic light that came down over the land like breath, like breathing.

"I remember the wild, dedicated lives of the birds that night and also the abandon with which a small herd of caribou crossed the Kokolik River to the northwest, the incident of only a few moments. They pranced through like wild mares, kicking up sheets of water across the evening sun and shaking it off on the far side like huge dogs, a bloom of spray that glittered in the air around them like grains of mica.

"I remember the press of light against my face. The explosive skitter of calves among grazing caribou. And the warm intensity of the eggs beneath these resolute birds. Until then, perhaps because the sun was shining in the very middle of the night, so out of tune with my own customary perception, I had never known how benign sunlight could be. How forgiving. How run through with compassion in a land that bore so eloquently the evidence of centuries of winter."

"Before we drove the few miles over to Deadhorse, the Prudhoe Bay airport, my host said he wanted me to see the rest of the Base Operations Building. A movie theater with tiered rows of plush red velour seats. Electronic game rooms. Wide-screen television alcoves. Pool tables. Weight-lifting room. Swimming pool. Squash courts. Running tracks. More television alcoves. Whirlpool treatment and massage. The temperatures in the different rooms are different perfectly. Everything is cushioned, carpeted, padded. There are no unpleasant sounds. No blemishes. You do not have to pay for anything. He shows me his rooms.

"Later we are standing at a railing, looking out through insulated glass at the blue evening on the tundra. I thank him for the tour. We have enjoyed each other. I marvel at the expense, at all the amenities that are offered. He is looking at the snow. 'Golden handcuffs.' That is all he says with his wry smile."

"Probably no other predator employs as many hunting strategies with one animal as the polar bear does with the ringed seal. It may take a half hour to patiently approach a seal resting on the edge of an ice floe, surfacing quietly to reconnoiter, then submerging again. A bear may drift toward a seal like an innocuous piece of ice; when it reaches the floe edge it explodes from the water and smacks the seal dead all in one motion. When it stalks seals over the ice, it flattens itself on its forequarters and slides along slowly on chest and forelegs, taking advantage of every piece of cover. It will scrape away the sea ice at a breathing hole until there is just a thin layer left, and then cover the ice with its body to cut off sunlight, so it looks to the seal below as if the thick crust of ice and snow are still present. It will build a snow wall to hide behind while it waits at an aglu [breathing hole]. And it will rise up suddenly in a resting seal's own aglu. . . .

"Just before it surfaces, the seal exhales, and the sight or sound of the bubbles alerts the bear. The seal rises headfirst up a cone-shaped tunnel to its breathing hole, which, on smooth ice, appears as a low mound. A small amount of water forced up ahead of the seal splashes out on the ice and freezes. (The seal keeps the tunnel open and the aglu from freezing over completely by scouring with its claws.) The bear must time its strike perfectly and move with exceptional speed. It usually strikes with one or both paws and follows so

21

quickly with its snout that if the smashing blow of its paws doesn't kill the seal, the impact of its snout will. 'Everything cooperates,' writes Van de Velde, '—paws, claws, snout, and teeth—to give a blow that is so rapid that the seal has hardly a chance of getting away.' "

Environmental Conservation
RAYMOND F. DASMANN
1984, John Wiley and Sons

It masquerades as a natural science textbook, but this comprehensive work is really a must-read for everyone interested in understanding the imperative question of our species' interdependence with the increasingly fragile planetary web of life. Ecologist Raymond Dasmann clearly feels this should be all of us and could have used the subtitle "How We Have Lived, Live Now, and Must Soon Change the Way We Live." Defining environmental conservation as "the use of the environment to sustain the greatest diversity of life while insuring for humanity the physical basis for continued well-being," he sets out to show the deep significance as well as radical potential of this fundamental subject. A section titled "Where Are We?" covers the present state of basic human requirements such as food, water, and energy. It also provides up-to-date background on biotic regions and bioregional boundaries that can fuel the growing sentiment for natural rather than arbitrary political borders. "Where Do We Want to Go?" describes the full horrific dimension of present ecological threats, including pesticides, pollution, and the wider problem of the "urban-industrial-technological trap" and then points toward remedial directions that suggest a completely different view of society. Eco-development (which means raising Third World living standards in ways that maintain natural systems and re-developing richer countries to be more ecologically sound) is an important one of them. So are appropriate technology, protection for Fourth World wild lands and first peoples, and bioregional reinhabitation.

A leading authority on worldwide conservation issues, Dasmann doesn't fail to cover his subject's basic working terminology (*ecosystems* and *biosphere*, for example) and essential technical concepts such as natural succession and the nitrogen cycle. But this completely revised fifth edition of a respected teaching sourcebook has taken on a fresh and urgent tone to stress the need for active engagement. He fully presents the field of study to which he has been a major contributor and never stints in letting you know what can be done about it.

—PETER BERG

"*Ecosystem people.* Hunter-gatherers have the longest history on earth and represent a way of life basic to humanity. . . . Contrary to earlier concepts, their lives are not a basic struggle for survival from which people gladly turn to agriculture. Rather it is a way of life preferred by its practitioners, and if we judge from the hunter-gatherer survivors, it is a way of living they do not willingly abandon.

"*Biosphere people.* The new biosphere culture draws on the resources of all ecosystems and is at the same time dependent on none. Destruction of resources in one area does not cause collapse of the culture since the global economy can shift to a different ecosystem and draw on its resources. It becomes possible therefore to carry out much more complete destruction of living resources than could have been achieved by the most profligate ecosystem people."

"Any way of life we pursue must be or be able to become ecologically sustainable. . . . There are various options to choose from, but there is no option that allows present trends to continue."

Man and Nature: Or, Physical Geography As Modified by Human Action
GEORGE PERKINS MARSH
1864; 1965, Harvard University Press

No national environmental organization bears his name or honors his memory. And few of today's most prominent activists are familiar with his life or work. Yet George Perkins Marsh (1801–1882) deserves to be regarded as this country's original environmentalist.

His single greatest accomplishment in a long and amazingly diverse and productive career was the publication in 1864 of the monumental book *Man and Nature*. In it Marsh upset the American myth of superabundance and warned, "The earth is fast becoming an unfit home for its noblest inhabitant, and another era of equal human crime and human improvidence . . . would reduce it to such a condition of impoverished productiveness, of shattered surface, of climatic excess, as to threaten the depravation, barbarism, and perhaps even extinction of the species."

He theorized, after years of scholarly research in parts of Asia Minor, northern Africa, Greece, Italy, and Alpine Europe, that deforestation, erosion, and land abandonment were contributory factors in the decline of past civilizations. And he concluded that "desolation, like that which has overwhelmed many once beautiful and fertile regions of Europe, awaits an important part of the territory of the United States . . . unless prompt measures are taken to check the action of destructive causes already in operation."

Much of the book, which quickly gained international reputation and was hailed by one reviewer as "one of the most useful and suggestive works ever published," is devoted to what was in Marsh's time the infant science of ecology. He explored the interrelationship of plant and animal communities, the succession of forests, the mechanics of dunes and barrier beaches, and the importance of aquifers.

The great lessons of the work, echoed by John Muir, Gifford Pinchot, Carl Schurz, Teddy Roosevelt, and others at the turn of the century, are that nature *does not* heal itself and that man has both the ability and responsibility to act in harmony with nature to prevent further destruction, and even to restore some past disturbances. It is this dual message of dissent and reform that set the agenda for all who followed.

Though *Man and Nature* is his most enduring work, Marsh was what his biographer, David Lowenthal, calls omnicompetent. He wrote a definitive work on the origin of the English language, wrote a Scandinavian grammar, translated Goethe, was a lawyer and served in the Vermont state legislature, headed a commission that designed the present Vermont State House and another to investigate logging practices and soil erosion, helped found the Smithsonian Institution while serving as the Vermont congressman, completed the final design of the Washington Monument, served as minister to Turkey, and finally in 1861 was appointed by President Lincoln to serve as the first American minister to the newly formed Kingdom of Italy, a post he held for twenty-one years. It was while serving in Italy that he wrote *Man and Nature*. He died in 1882 in Vallombrosa in the Apennines, while visiting a friend who had established a school of forestry, and was buried in the Protestant cemetery in Rome.

—PETER BORRELLI

"Man is everywhere a disturbing agent. Wherever he plants his foot, the harmonies of nature are turned to discords."

"Our inability to assign definite values to [the] causes of the disturbance of natural arrangements is not a reason for ignoring the existence of such causes in any general view of the relations between man and nature, and we are never justified in assuming a force to be insignificant because its measure is unknown, or even because no physical effect can now be traced to it as its origin."

"It is a legal maxim that 'the law concerneth not itself with trifles,' . . . but in the vocabulary of nature, little and great are terms of comparison only; she knows no trifles, and her laws are as inflexible in dealing with an atom as with a continent or a planet."

A Sand County Almanac and Sketches Here and There

ALDO LEOPOLD

1949, Oxford University Press

Aldo Leopold is routinely called a modern American prophet, and *A Sand County Almanac* has become the "bible" of the contemporary environmental movement. In *Sand County* Leopold provides a fresh, scientifically informed vision of a practicable, harmonious human–environment relationship for the postindustrial age. The author's eminence as a professional conservationist lends his book credibility and authority. The elegance and simplicity of his prose give his message beauty, power, and accessibility.

Even though at first glance it seems to be a miscellaneous hodge-podge, Leopold's classic is in fact wonderfully unified and tightly organized. Granted, many of its forty-some-odd seasonal vignettes, geographically scattered sketches, and reflective essays on wilderness, wildlife, aesthetics, and ethics were written and separately published over a period spanning three decades. However, Leopold substantially revised some, blended others, and added newly written material to create his much celebrated chef d'oeuvre. Thus the finished book is more like a literary analogue of an ecosystem than a volume of collected essays. While each element remains genetically distinct and retains its separate identity, all are bound into a seamless whole by the concept of a biotic community, the thematic thread running through every piece of *Sand County*.

"The Land Ethic," which stands at the end of the book, is its philosophical climax and consummation. In it Leopold affirms Darwin's earlier account of the origin and evolution of our social instincts and moral sentiments. And he points out that ecology has more recently informed us that we are plain members and citizens of biotic communities, no less than of human societies. Therefore, a land ethic, he simply and straightforwardly argues, "implies respect for . . . fellow-members and also respect for the community as such."

—J. BAIRD CALLICOTT

"The 'key-log' which must be moved to release the evolutionary process for an ethic is simply this: quit thinking about decent land-use as solely an economic problem. Examine each question in terms of what is ethically and esthetically right, as well as what is economically expedient. A thing is right when it tends to preserve the integrity, stability, and beauty of the biotic community. It is wrong when it tends otherwise."

"One of the penalties of an ecological education is that one lives alone in a world of wounds. Much of the damage inflicted on land is quite invisible to laymen. An ecologist must either harden his shell and make believe that the consequences of science are none of his business, or he must be the doctor who sees the marks of death in a community that believes itself well and does not want to be told otherwise."

"We reached the old wolf in time to watch a fierce green fire dying in her eyes. I realized then, and have known ever since, that there was something new to me in those eyes—something known only to her and to the mountain. I was young then, and full of trigger-itch; I thought that because fewer wolves meant more deer, that no wolves would mean hunters' paradise. But after seeing the green fire die, I sensed that neither the wolf nor the mountain agreed with such a view.

*   *   *

"Since then I have lived to see state after state extirpate its wolves. I have watched the face of many a newly wolfless mountain, and seen the south-facing slopes wrinkle with a maze of new deer trails. I have seen every edible bush and seedling browsed, first to anaemic desuetude, and then to death. I have seen every edible tree defoliated to the height of a saddle-horn. Such a mountain looks as if someone had given God a new pruning shears, and forbidden Him all other exercise. In the end the starved bones of the hoped-for deer herd, dead of its own too-much, bleach with the bones of the dead sage, or molder under the high-lined junipers."

Soil and Civilization
EDWARD HYAMS
1952; 1976, Harper & Row, out of print

Long-term degradation of soil has often not been discernible to
people who live on the land and has entered the minds of only a very
small percentage of those who live in the cities. The current docu-
mentation of soil erosion, decay of rural communities, and urbanization of land on a national scale has not moved the populace to give
priority to such problems. Instead, the agricultural agenda has been
mostly concerned with controlling the level and export of the abundant food production which has been made possible by modern
technology and a dowry of good soils. The dangerous (not so much
to us as to our descendants) complacency brought about by this
veneer of success makes *Soil and Civilization* an important reminder
of our dependence on the nurturing of rural communities and good
soils. Hyams examines the historical relationship of early civilizations to the soils on which they practiced agriculture. On the fertile
alluvial soils of river systems such as the Nile, the Euphrates-Tigris,
the Indus, and the Hwang-Ho, he regards humans as soil parasites
because the fertility of the soils was maintained by the constant
deposition of silt by the rivers, not by the effort of humans. Examples
are then given of humans as a disease of soils, such as the rise of
Athens at the expense of the destruction of the soils of Attica. After
discussing the marginal cases of Eurasia, China, and India, Hyams
expounds on humans as soil makers in the Inca Empire and in
Europe. Throughout the book and at the conclusion, he elaborates
on the state of mind and spirit needed for humans to regard themselves as members of the soil community.

—MARTY BENDER

"Urban Western man, accustomed to being flattered as uniquely
powerful in material matters, by newspapers whose advertisers need
the goodwill of their readers, are prone to ignore the massive
achievement of past civilizations in the manipulation of matter. It is

commonly thought that mechanical science has conferred upon us powers of manipulation different in kind from those possessed by our predecessors. . . . It should be realized that there is no significant difference between a flint or bronze digging tool and a bulldozer. Almost unlimited slave power can achieve quite as much as mechanical power. The Carthaginians applied to the exploitation of their soil the device of plantation-slavery upon a very large scale.

"The device is one of the most vicious means of expressing power yet tried. It was the ruin of Rome, who imitated it. And it was the destruction of Carthage when she came into conflict with a Rome which had not *yet* imitated it upon a serious scale. The truth is as Goldsmith saw it: the insight of the poet is, as usual, more immediately effective in arriving at the truth than the careful experiments of science and economy:

> *Ill fares the land to hastening ills a prey,*
> *When wealth accumulates and men decay.*
> *Princes and peers may flourish and may fade,*
> *A breath can make them, as a breath has made.*
> *But a proud peasantry, its country's pride,*
> *When once destroyed can never be supplied.*"

"*Tools, Techniques, and States of Mind and Spirit*

"There is no peculiar merit in ancient things, but there is merit in integrity, and integrity entails the keeping together of the parts of any whole, and if those parts are scattered throughout time, then the maintenance of integrity entails a knowledge, a memory, of ancient things. . . . To think, feel or act as though the past is done with, is equivalent to believing that a railway station through which our train has just passed, only existed for as long as our train was in it. A community which ignores or repudiates its origins, in its present acts, is no more whole and healthy than a man who has lost his memory. One of the conditions for the achievement of full human status by man is that he should 'remember' every detail of his past; and this is the importance of all the arts and sciences which recreate the past in our consciousness. . . . Everything that man has ever done in his relationship with soil is significant for what he now does, and agricultural man can no more safely ignore his past than architectural man can ignore the Gothic cathedrals or the Baroque palaces, or a mathematician afford to know nothing of Newton. Man, being an organism living on organisms, his works have organic attributes, and the work man does grows out of the work he has done in the past."

The Unsettling of America: Culture and Agriculture
WENDELL BERRY
1977; 1986, Sierra Club Books

This 1977 book is a classic critique of modern American agriculture and culture. Many people working to promote sustainable agriculture credit the reading of *The Unsettling* for their awakening to the devastating impact of agribusiness upon the land and rural communities. College students, from both rural and urban backgrounds, have become committed to reform after they have read *The Unsettling* as a class assignment.

The book is a compilation of nine separate essays, some of which were first written for publication elsewhere. The first essay, bearing the title of the book, introduces Wendell Berry's outrage at agricultural policies, like those of Earl Butz, former secretary of agriculture in the Nixon administration, who referred to "food as a weapon." Such policies encourage exploitation instead of nurture, overuse of the land in pursuit of high production and efficiency, and the demise of family farms and rural communities, the "unsettling of America."

Although Berry values natural places unspoiled by humans for reasons he lists, he wants conservationists to be aware of how the Earth *must* be used in their behalf. "Kindly use" of land, in particular farm land, depends upon "intimate knowledge," which large-scale, industrial agriculture cannot afford. Consumers, even conservationists, want cheap, convenient food. Cheapness and convenience come at the expense of health, both for people and the land.

The book is infinitely quotable. A poet and novelist as well as an essayist, Wendell Berry uses the English language with love and care, and the reader marvels at his clear expression of the complex. He writes as a moralist about character and culture, soil, the use of energy, the future of agriculture, land grant colleges, and human sexuality. In this last category, a patriarchal perspective pervades, though he seems unaware of his condescension toward women. But this should not stop the reader from appreciating the overall wisdom and beauty of the book. It is a landmark, essential reading for anyone trying to understand the components of a sustainable society.

—DANA JACKSON

"I had stopped once and talked a while with the old man. He was busy fixing a fence at the time, and though he received me courteously enough, he did not permit himself to be much interrupted. I told him that I admired his farm. He thanked me, but without enthusiasm, obviously having spent little time yearning to be complimented by strangers. I said his team of horses looked like a good one. He said that they did very well.

"One morning after I had learned of his death, I stopped at the farm again—in his honor, maybe, or in honor of my own sense of loss. It was a gray, wintery day. The place looked and felt forgotten. It had gone out of mind. Absence was in it like a force. The barn was closed, empty, the doors tied shut by someone who did not intend to come back very soon. Peering through a crack, I found that I was looking into a milking room with homemade wooden stanchions, unused for years. I knew why: it had become impossible to be a *small* dairyman. I spent some time looking at the old man's horse-drawn equipment. Some antique collector had taken the metal seats off several of the machines; these had become bar stools, perhaps, in somebody's suburban ranch house. For the rest apparently nobody now had a use. Examining the pieces of equipment, I saw that they were nearly completely worn out, patched and wired together like the fences and buildings, made to do—the forlorn tools of a man who had heirs, but no successors."

"Values may be corrupted or abolished in only one discipline at the start, but the damage must sooner or later spread to all; it can no more be confined than air pollution. If we corrupt agriculture we corrupt culture, for in nature and within certain invariable social necessities we are one body, and what afflicts the hand will afflict the brain."

BILL MOLLISON, WES JACKSON, AND MASANOBU FUKUOKA

Half the farming regions of humanity's first civilizations are now deserts. And the condition of Africa is warning that there is something unsustainable about how agriculture raises food. Topsoil to raise food is a restorable resource when it is there in the first place. But water and wind erosion carry it away to the bottom of rivers and oceans. And the natural processes by which soil is made take so many thousand years that it is practically nonrenewable.

Today three men prophecy a new, sustainable way of food growing: the Australian Bill Mollison, the American Wes Jackson, and the Japanese Masanobu Fukuoka. Their answer to the destructive character of civilized agriculture: Stop plowing. Leaving plant growth undisturbed year after year built the topsoils that we have. Mollison, Jackson, and Fukuoka are developing a perennial food growing that is never plowed. Not that all three are proposing exactly the same thing. (There is a wonderful photograph of them clowning as Buddhist monkeys: see no evil, hear no evil, speak no evil.) But elements of the vision of each make up a powerful new realm of food growing that is radically other than the agriculture of the last ten thousand years.

Mollison, whose ideas are elaborated in *Permaculture: A Practical Guide for a Sustainable Future,* leads us beyond traditional yards, fields, pastures, and woodlots into a new world of the "permaculture" design of landscape: a permanent horticulture and agriculture, which enriches rather than destroys. He teaches us to take the lay of the land, the flow of different temperatures of air, the channeling of water, configurations and relations of trees, microclimates in a planned landscape, all imagination about patterning of biology, to design an architecture of living things to sustain us. Mollison is most concerned with reclamation of desert and marginal lands rather than change of agriculture itself.

Jackson lives on the great American prairie, where soil erosion is severe. Nationally, we lose between three and twenty tons of topsoil *per acre per year;* so you can see why Jackson is confident plowing will not last forever. He works to develop perennial food grains that are harvested but never plowed. Depending on how the genetics of such changes are handled, this evolution beyond plowed agriculture would not be unlike the ancient selection and breeding that created all our food plants.

Fukuoka offers another, more radical part of the picture. Beginning with his *One Straw Revolution* in 1975, he has been developing his own method of unplowed "natural" farming. He sows and reaps traditional seeds by small labor-intensive methods, without the machinery that Jackson's vision involves. In his latest book, *The Natural Way of Farming*, Fukuoka gives details of the sequences of hand seeding and harvesting over a year that are crucial to his method. In the transformed farming that Fukuoka offers, almost everyone in society will be involved in food growing on their own quarter-acre plots.

In Mollison and Fukuoka's books, you are given plenty of things you can do in your own backyard, and in Jackson's books *New Roots for Agriculture* and *Altars of Unhewn Stone*, plenty of reasons why you should start now.

—PHIL HOLLIDAY

Permaculture: A Practical Guide for a Sustainable Future
BILL MOLLISON
1990, Island Press

New Roots for Agriculture
WES JACKSON
1980, University of Nebraska Press

The Natural Way of Farming
MASANOBU FUKUOKA
1987, Japan Publications

"The philosophy behind permaculture is one of working with, rather than against, nature; of protracted and thoughtful observation rather than protracted and thoughtless action; of looking at systems in all their functions, rather than asking only one yield of them. . . ."

"There are so many opportunities to *create* systems that work from the elements and technologies that exist. Perhaps we should do nothing else for the next century but apply our knowledge. We already know how to build, maintain, and inhabit sustainable systems. Every essential problem is solved, but in the everyday life of people this is hardly apparent."

"Permaculture as a design system contains nothing new. It arranges what was always there in a different way, so that it works to conserve energy or to generate more energy than it consumes. What is novel, and often overlooked, is that *any* system of total common-sense design for human communities is revolutionary!"

—*Permaculture*

"Those interested in the long-term health of the land need only stand on the edge of a stream after a rain and watch a plasma boil

34

and turn in the powerful current below and then realize that the vigorous production of our fields is, unfortunately, temporary. Since we initiated the split with nature some 10,000 years ago by embracing enterprise in food production, we have yet to develop an agriculture as sustainable as the nature we destroy."

"And, we might add, among those works would be a new agriculture closer to our 'original relation to the universe.' Our old agriculture is rooted in a tradition that is basically ruinous and consequently a tradition from which we must extricate ourselves. We will have to employ the very best practices of conservation and agriculture that are with us today. We will have to continue to support those selfless people and organizations dedicated to the principles and practices of conservation.

"But in the long run, a new agriculture will be necessary. I believe it can and must be done and that when it is underway, it will rank as one of the greatest human ventures of all time, requiring more imagination and truly joyful participation than any of us can now imagine."

—New Roots for Agriculture

"Flood a field with water, stir it up with a plow, and the ground will set as hard as plaster. If the soil dries and hardens, then it must be plowed each year to soften it. All we are doing is creating the conditions that make a plow useful, then rejoicing at the utility of our tool. No plant on the face of the earth is so weak as to germinate only in plowed soil. Man has no need to plow and turn the earth, for microorganisms and small animals act as nature's tillers."

"This is the purpose of quarter-acre farming for all the people of the land. If people have a change of heart, they will not need vast green fields to achieve this rebirth; it will be enough for them to work small fields. Our world has fallen into a state of chaos because man, led astray by an embarrassment of knowledge, has engaged in futile labors. The road back to the land, back to the bosom of a pure, innocent nature still remains open to us all."

—The Natural Way of Farming

The Encyclopedia of Organic Gardening
BY THE STAFF OF *ORGANIC GARDENING* MAGAZINE
1978, Rodale Press

Since gardening's return to popularity in the early 1970s, gardeners have evolved from growing vegetables for self-sufficiency to adding herbs and flowers and, more recently, to attempting full-blown landscapes with naturalized "environments" and formal gardens of exotic plants.

These changes have been documented and encouraged by a golden age of horticultural publishing. Both American and British publishing houses are producing breathtakingly beautiful books that both inspire and inform. They address the mysterious and loving relationship we humans can have with plants by appealing to our quest for knowledge, our instincts for husbandry, and our delight in the beauty of living things.

My favorite way to learn about growing plants is first to look at the garden, deciding what needs doing and how it might be done. Then I consult the one gardening book in my collection that rarely sits on the shelf, *The Encyclopedia of Organic Gardening* from Rodale Press.

If the irises were small this year, they probably need dividing; *The Encyclopedia* tells me how. If the vegetables need mulch, but I don't have grass clippings and can't cram a bale of straw into the trunk of my car, *The Encyclopedia* lists and describes the use of a variety of alternatives, such as old leaves shredded with a lawn mower. It explains the cultivation practices for most commonly grown food crops, grasses, trees, flowers, and shrubs. There is also a wealth of information on soils and composting, true to Rodale's traditional emphasis on gardening without chemicals.

Like that kitchen classic *The Joy of Cooking, The Encyclopedia* is periodically updated and offers both how-to recipes and the means for turning our own ideas into action.

(Note: The Garden Book Club is a good way to enjoy this golden age of gardening books. Its monthly catalog describes dozens of the latest American and British releases. The Garden Book Club, 250 West Fifty-seventh Street, New York, NY 10107.)

—ROSEMARY MENNINGER

"Intensive Gardening

"Although ideal for pocket-sized gardens, intensive gardening was not originally designed for that purpose. Instead, it was designed to increase yields and improve their quality without a large capital investment. No farm machinery is needed to raise and harvest crops that produce up to four times as much as crops grown under conventional methods.

"Though intensive gardening has been practiced in one form or another for many centuries, the basic methods remain the same: the use of raised beds; double digging; organic fertilizers and compost; nonlinear, intensive planting; and companion planting."

"Parsnip (Pastinaca sativa)

"In the East and North, this root vegetable can be left in the ground all winter and dug up as needed for cooking. Freezing seems to improve the texture and gives parsnips a sweeter, more delicate taste. In southern and western states where winters are mild, parsnips should be planted in fall and grown for a winter crop, because spring planting extends the warm growing season too long, making the parsnips woody and tasteless."

"Chopped bark is perhaps the most attractive mulching material available, and is used in large quantities in parks and professionally landscaped homes. Redwood, red cedar and cypress barks are relatively impervious to decay and will last for years. Oak and pine barks are high in acid and should have an extra sprinkling of lime put on the soil under them, or be used with acid-loving plants."

Diet for a Small Planet (tenth anniversary edition)
FRANCES MOORE LAPPÉ
1971; 1985, Ballantine Books

Food First: Beyond the Myth of Scarcity
FRANCES M. LAPPÉ AND JOSEPH COLLINS
WITH CARY FOWLER
1979; 1981, Ballantine Books

Diet for a New America: How Your Food Choices Affect Your Health, Happiness, and the Future of Life on Earth
JOHN ROBBINS
1987, Stillpoint Publishing

By the time the tenth anniversary edition appeared in 1982, *Diet for a Small Planet*—published originally in 1971—had sold close to two million copies and popularized the idea that animals were "protein factories in reverse." Branding meat production as wasteful because "the average ratio of all U.S. livestock is 7 pounds of grain and soy fed to produce 1 pound of edible food," Lappé develops two interrelated premises: (1) the grain-fed meat diet is a major factor in the destruction of the environment and (2) protein sources from plants and dairy products can replace meat protein with no ill effects on humans.

[Editor's note: Lappé and Collins in *Food First* provide a clearly written and exhaustively researched indictment of multinational agribusiness as perpetuator—and perpetrator—of colonialist practices that cause rural impoverishment and world hunger. The scope of the analysis in *Food First* is a global, necessary, and logical outgrowth of the concerns addressed in *Diet for a Small Planet*.]

While Lappé is the founding mother of the analysis of the ecological impact of meat eating, John Robbins, in *Diet for a New America*, examines in a popular, accessible style the ethics of eating meat, dairy products, and eggs, as well as issues such as the association between meat production and the disappearance of the rain forests. Both Lappé and Robbins view a plant-based diet as a healthier diet and one the Earth's resources can sustain. Their books empower individuals to see how they can challenge global patterns of environmental abuse by beginning with their own food choices.

—CAROL J. ADAMS

"Our grain-fed meat diet not only wastes our resources but helps destroy them too. Most people think of our food-producing resources, soil and water, as renewable, so how can they be destroyed? The answer is that because our production system encourages farmers to continually increase their output, the natural cycle of renewal is undermined. . . . Here are a few facts to give you some sense of the threats to our long-term food security:
• *Soil erosion.* Corn and soybeans, the country's major animal feed crops, are linked to greater topsoil erosion than any other crops. In some areas topsoil losses are greater now than during the Dust Bowl era. At current rates, the loss of topsoil threatens the productivity of vital farmland within our lifetime.
• *Import dependency.* Corn alone uses about 40 percent of our major fertilizers. U.S. agriculture has become increasingly dependent on imported fertilizer, which now accounts for 20 percent of our ammonia fertilizer and 65 percent of our potash fertilizer. And even though the United States is the world's leading producer of phosphates for fertilizer, at current rates of use we will be importing phosphates, too, in just 20 years."

—*Diet for a Small Planet*

"Diet-style changes could not only halt the process of deforestation, they could actually reverse it. Of the 260 million acres of American forest that have been converted into land now used to produce the standard American high-fat low-fiber diet-style, well over 200 million acres could be returned to forest if Americans were to stop raising food to feed livestock, and instead raise food directly for people. Indeed, so direct is the relationship between meat production and deforestation that Cornell economist David Fields and his associate Robin Hur estimate that for every person who switches to a pure vegetarian diet, an acre of trees is spared every year. A lacto-ovo vegetarian diet-style is also helpful, particularly if dairy and egg product consumption is low."

"Recent studies indicate that of all the toxic chemical residues in the American diet, almost all, 95% to 99%, comes from meat, fish, dairy products and eggs. If you want to include pesticides in your diet, these are the foods to eat. Fortunately, you can overwhelmingly reduce your intake of these poisons by eating lower on the food chain, and not choosing foods of animal origin."

"Some people feel that eating 'organically raised' beef and poultry is a good way to limit their intake of pesticides. It is important to realize, though, that while meat products labeled 'natural' or 'organic' may be better than the typical factory farm commercial products, they still will include the concentrated toxins from all the foods the livestock ate. These lethal chemicals accumulate in the fatty tissues of animals in much greater concentrations than are found in fruits and vegetables."

—Diet for a New America

THE EHRLICH OPUS

I know I'm not the only person whose life took a new turn—toward the ecology movement—in consequence of a head-on encounter with Paul Ehrlich's *The Population Bomb*, which effectively announced the problem of overpopulation to the American public. Certainly this best-seller and its author were among the principal catalysts of the ecology movement of the early seventies.

Sensational as *The Population Bomb* was, its author is no one-trick pony. Ehrlich is a distinguished scientist who continues to make significant contributions to evolutionary theory. An avid field biologist who has caught and released checkerspot butterflies too numerous to count, Ehrlich is Bing Professor of Population Studies at Stanford University.

Paul Ehrlich is author, coauthor (frequently with his wife Anne, his colleague in research, writing, and activism since the mid-fifties), or editor of hundreds of columns and articles and twenty-one books. These range from respected texts to anthologies to accessible treatments of uncomfortable social issues, such as race and immigration, and of nuclear winter, an even grimmer prospect than terminal overpopulation. His creative and intellectual energy is prodigious. Vitalized by his love of and for his subjects, Ehrlich's endless delight in biology is revealed in his recent (1986) book *The Machinery of Nature,* a deceptively conversational, cannily organized exposition of the evolutionary approach to ecology.

The most recent joint production of the Ehrlich's alarm factory is *The Population Explosion,* which asserts, anew for the 90s, that overpopulation is a driving force in the current array of global environmental problems, just as *The Population Bomb* did in its day. It is a lack of progress report, bluntly describing humanity's failure to take adequate measures to curtail population growth and the misery it produces. Substantially documented and ideologically liberal, *The Population Explosion* is lively, must reading that somehow manages to be positive in encouraging the activists and policymakers who may heed its warnings.

Scrappy and sardonic, Paul Ehrlich is a scientist-citizen-teacher of real stature and staying power. His contributions, and Anne Ehrlich's, to the essential knowledge base of the ecology movement have been great. And, if the preceding thirty years are any indicator, they will continue to be.

—STEPHANIE MILLS

The Population Bomb
1968; 1971, Ballantine Books

The Population Explosion
1990, Simon & Schuster

The Machinery of Nature
1986, Simon & Schuster

"Our own species, *Homo sapiens*, evolved a few hundred thousand years ago. Some ten thousand years ago, when agriculture was invented, probably no more than five million people inhabited Earth—fewer than now live in the San Francisco Bay Area. Even at the time of Christ, two thousand years ago, the entire human population was roughly the size of the population of the United States today; by 1650 there were only 500 million people, and in 1850 only a little over a billion. Since there are now well past 5 billion people, the vast majority of the population explosion has taken place in less than a tenth of one percent of the history of *Homo sapiens*.

"This is a remarkable change in the abundance of a single species. After an unhurried pace of growth over most of our history, expansion of the population accelerated during the Industrial Revolution and really shot up after 1950. Since mid-century, the human population has been growing at annual rates ranging from about 1.7 to 2.1 percent per year, doubling in forty years or less. . . .

"But even the highest growth rates are still *slow-motion changes compared to events we easily notice and react to.* . . . In four years, the world population expands only a little more than 7 percent. Who could notice that? Precipitous as the population explosion has been in historical terms, it is occurring at a snail's pace in an individual's perception. It is not an event, it is a trend that must be analyzed in order for its significance to be appreciated."

—The Population Explosion

"Coevolution does often seem to go in this direction; in many cases, parasites attacking host populations that have long been exposed to them seem less lethal. In over four million square miles of middle Africa, newly introduced cattle quickly sicken and die. They are victims of nagana, a sleeping sickness of cattle caused by a trypanosome (a protozoan that, like malaria, invades the blood). This protozoan is closely related to the agent that causes human sleeping sickness and, like the latter, is carried by tsetse flies. Varieties of cattle that have been in regions infected with nagana for a long time, such as the small, humpless n'Dama cattle in West Africa, show some resistance. Native antelopes, on the other hand, which have been coevolving with the trypanosomes for millennia, harbor them with no apparent ill effects. The impact of parasite on host has clearly been softened by the long association."

"Above all, the fates of the scientific enterprise and *Homo sapiens* now seem to be inextricably connected. If both survive and prosper, I believe it will be for two reasons. One will be that accurate information about science, what it has discovered, and how it works, has become much more widely disseminated. The second will be that more and more people will learn there are other sources of knowledge besides science—that, for example, science cannot provide sure solutions to many ethical problems. It can often clearly delimit ethical alternatives, but the wisdom to select the right one will, in many cases, have to be found elsewhere."

—The Machinery of Nature

Biodiversity
EDITED BY E. O. WILSON
1988, National Academy Press

From the temperate rain forests of Oregon to their analogues in tropical Amazonia, the exponentially increasing human population is eliminating genes, species, and ecosystems—biological diversity— at a rate unknown for sixty-five million years. An act of unequaled

43

enormity in itself, it threatens our existence, for people depend on living things for every bite of food, every drink of water, every breath of air, for myriad products and services. That is one basic message of this compendium edited by the justly renowned ecologist Ed Wilson.

But there is another message: that we can slow the tide of extinction and ecosystem destruction. By understanding both biological systems and the economic and social systems that are consuming them, we can sustain life on Earth. It will not be easy, but these essays lay the groundwork.

Ed Wilson, Peter Raven, Norman Myers, and Carleton Ray sketch out the dimensions of the crisis. Norman Farnsworth, Hugh Iltis, and Mark Plotkin provide examples of what we stand to lose. Jerry Franklin shows how maintaining structure—especially dead trees—is vital to temperate forest species. Dan Janzen describes steps for restoring tropical dry forests. Chris Uhl explains the inherent fragility and resilience in Amazonian forests. Joy Zedler explores pitfalls in efforts to restore degraded salt marshes. Mike Robinson, Tom Lovejoy, Bob Jenkins, and Betsy Dresser provide unique insights on ways we can conserve biological diversity.

Although it neglects some important topics and its References layout is terrible, *Biodiversity*'s diverse, authoritative perspectives make it a must for anyone interested in conservation.

—ELLIOTT A. NORSE

"The diversity of life forms, so numerous that we have yet to identify most of them, is the greatest wonder of this planet."

—E. O. WILSON

"If we cannot act as responsible stewards in our own backyards, the long-term prospects for biological diversity in the rest of this planet are grim indeed."

—DENNIS D. MURPHY

"When economists from the United States and other developed countries urge developing countries to preserve their biological resources, there is a certain awkwardness. It is like an aging rake urging chastity on a young man: the advice is certainly based on a wealth of experience, but it may not be entirely persuasive."

—W. MICHAEL HANEMANN

"Biotic diversity is not linked to the distribution of elephants, rhinos, and other so-called charismatic megaherbivores. The massive investment in conservation campaigns does more for the souls of the donors and the egos of the elephant experts than it does for biotic diversity, which is centered on less exciting communities of montane forests, Mediterranean heathlands, wetlands, lakes and rivers."

—BRIAN J. HUNTLEY

". . . technology is not a panacea for extinction. It is a palliative— a topical treatment with which to buy time, to preserve options for a few populations and species judged of special value."

—WILLIAM CONWAY

BIODIVERSITY AND THE ANCIENT FORESTS OF THE PACIFIC NORTHWEST

As the conquering of North America by white Anglo settlers has rapidly proceeded, most have been willing to sacrifice the richness and diversity of a wild continent for a standard of living never seen before in history. In our wake we have left mere shreds of what were once vast bison herds, old-growth broad-leaved forests, native grasslands, clouds of waterfowl, and a host of ghost species, now extinct. From an ecologist's view, this occupation seems to have as its ultimate goal the makeover of wild nature into a natural zoo for our benefit.

But as human appropriation of native habitats has progressed, so has our understanding of the very ecosystems we seem bent on decimating.

If the number of recently published books is any indication, we may finally be awakening to the profound implications of the erosion of biological diversity. Each of the following three books focuses on the last remnants of one of the world's greatest forests: the old-growth Douglas fir forests of the Pacific Northwest. This is as it should be. Nowhere else in the United States is there more at stake: An entrenched timber economy threatens the remaining 13 percent of a magnificent endangered ecosystem.

This view of the old-growth debate stands at odds with the more conventional wisdom expressed in a recent *Washington Post* review of Keith Ervin's *Fragile Majesty*. The *Post* believes the controversy to be of "modest magnitude and limited geography," a view which author Ervin does not share. *Fragile Majesty* is his evenhanded, journalistic account of the "premier issue of public land management in the west." This book is the single best telling of the multitude of stories of native peoples, loggers, biologists, Earth First!ers, and Forest Service managers that weave together into the old-growth tale.

Ecologist Elliott Norse, in *Ancient Forests of the Pacific Northwest,* also makes a claim for the global significance of the ancient forest. Geared to the layperson, Norse's book provides a compelling ecological overview of the last stand of old-growth ecosystems. Along the way he dispels what may be humanity's premier ecological fallacy: that we can have unrestrained growth without regard to protecting biological diversity. Like Ervin, Norse also provides recommendations for sustaining both ancient forests *and* forestry in the Pacific Northwest.

If protecting biodiversity requires an intimate understanding of ancient forests, then *Forest Primeval,* by biologist Chris Maser, brings us close to wisdom. Beginning in A.D. 988 Maser traces the life of a single old-growth stand from birth to maturity. Along the way you meet the animals and plants whose lives help define what ancient forests are: fungi, carpenter ants, flying squirrels, cougars, and more. After reading *Forest Primeval,* one realizes that the notorious spotted owl is but one ancient forest dweller among many.

Forest Primeval is a natural history tour de force and suggests that the conflict over old growth is as value laden as it is a matter of ecology and economics.

For all the wealth of information provided by these timely books, Maser's point stands clear. Whether it is the ancient forests in the Pacific Northwest or those in Brazil that are considered, the biodiversity crisis is not about numbers. It is finally a test of our values, our ability to respect the limits placed on those who dwell on a finite planet, and our wisdom to "let beings be."

—R. EDWARD GRUMBINE

Fragile Majesty: The Battle for America's Last Great Forest
KEITH ERVIN
1989, Mountaineers Books

Ancient Forests of the Pacific Northwest
ELLIOTT NORSE
1989, Island Press

Forest Primeval: The Natural History of an Ancient Forest
CHRIS MASER
1989, Sierra Club Books

"If the objective is preserving old growth as an ecosystem, scientists argue, then the emphasis ought to be on the ecosystem, not a handful of its occupants."

"While timber executives proclaim a crisis on public lands and berate the Forest Service for jeopardizing 5,000 jobs, these executives offer no apologies for the loss of 55,000 jobs due to depletion of the industrial forests and automation in the mills."

"Roughly five-sixths of the forestland of western Washington and Oregon has been transformed from a natural ecosystem to a landscape controlled by humans."

—*Fragile Majesty*

"Loggers and environmentalists might seem to have little in common, but many share at least one thing: acceptance of the 'either/or' fallacy: Either we have biological diversity or timber production."

"Whether you care about curing cancer, reaching your timber quota, drinking clear water, bagging a buck, or reveling in hushed green serenity, you have to maintain biological diversity."

"Because biological diversity is the source of **all** we value from Westside forests, maintaining it is the **highest** priority for forestland management."

—*Ancient Forests of the Pacific Northwest*

"As we destroy the forests of the world, for whatever rational reason, we are, as a global society, simultaneously destroying our historical roots and grossly impairing our spiritual well-being."

"What we are doing to the forests of the world is but a reflection of what we are doing to ourselves and to one another."

—*Forest Primeval*

In the Rainforest: Report from a Strange, Beautiful, Imperiled World
CATHERINE CAUFIELD
1985, University of Chicago Press

Lessons of the Rainforest
EDITED BY SUZANNE HEAD AND ROBERT HEINZMAN
1990, Sierra Club Books

It's a pretty straightforward proposition: If we destroy our rain-forests, and the millions of species of insects, birds, plants, and trees that live in them, Earth will be uninhabitable, at least by people as we know us.

And we are, of course, killing the forests, at a horrendous rate. That much is beyond debate. The good news is that we can do something about it.

As with any self-cure, the first step is education. When Catherine Caufield's *In the Rainforest* appeared in 1985, most laypeople hardly knew what a rainforest was. That's changed, mightily, and due in no small part to Caufield's book (it's been translated into thirteen languages). Above all, she's a good reporter. In a hard-nosed, fact-filled, highly readable style, she takes you into rainforests around the world—the Amazon, Indonesia, Malaysia, Africa. You get history, you get science, you get a close dissection of the colos-sally twisted workings of big business and the so-called development institutions that are the primary funders of rainforest destruction.

In the years since *In the Rainforest* was published, the numbers have changed—they're worse—but the paradigms have not. Mean-while, our overall fund of rainforest knowledge has increased expo-nentially. *Lessons of the Rainforest* will bring you up to date. It's a tool chest of articles by twenty-four rainforest experts, including Anne and Paul Ehrlich, Norman Myers, Donald Perry, and Jason Clay.

Lesson number one from the rainforest, as presented here, is:

Everything is connected. We can't save the rainforests if we don't change how we live.

We change first as individuals, as consumers, by understanding how intricately our daily lives—starting with that wake-up cup of rainforest-bred coffee—intertwine with the rainforest itself, a world that at first thought might seem dark, gooey, and terribly remote.

Then we change as a society. One of the hundreds of well-made points in *Lessons* is that the U.S. tax dollar, via the World Bank and such other highly secretive investment bodies as the Inter-American Development Bank, is still the principal destroyer of rainforests worldwide (though Japan is challenging for that position). But in 1989, in direct response to pressure from outraged U.S. citizens, these institutions finally began to awaken to the environmental havoc they've caused.

Lessons of the Rainforest will teach you about the banks, about the forests themselves, about the people and plants and animals who live in them, and about what you can do to save what's left. And that may be the most important lesson: The rainforests can be saved, but only if we as individuals *act*. Start by reading these books.

—JOE KANE

"*A Numbers Game or Survival*

". . . According to the Rainforest Action Network, the United States imports about $2.2 billion worth of tropical hardwoods each year, over one-fourth of the $8 billion annual trade in tropical timbers. For every foot of tropical plywood or paneling we buy, much more forest is destroyed in the logging process.

"However, tropical hardwoods—teak, mahogany, rosewood, meranti, ebony, lauan, and others—are luxury woods much prized by U.S. consumers. Furniture, picture frames, tableware, and paneling made of these woods are status symbols for homes and offices, signs of wealth and power; so are boats made of teak. And what are status symbols, if not projections of our self-esteem and position in the world? Perhaps if we were more committed to our bodies and our planet, we would rely less on the symbols of conspicuous consumption, and therefore consume less.

"Our consumption of tropical wood and wood products—in effect our consumption of tropical rainforests—is something that we in the North definitely can change. But it will require a change within ourselves, a change of attitude and perspective."

—*Lessons of the Rainforest*

State of the World
Annual, Worldwatch Institute

Worldwatch Paper
Series, Worldwatch Institute

Worldwatch Magazine
Worldwatch Institute

World Resources
Annual, World Resources Institute

For up-to-date, readable statistics and analysis on complex global environmental and resource trends, you'd do well to start with the publications of Worldwatch and World Resources Institute (WRI). Worldwatch, formed in 1974 by Lester Brown, has been concerned with identifying and analyzing emerging global problems and trends, such as soil erosion and ozone depletion, and bringing these issues to the attention of the general public and opinion leaders. The institute has published annual *State of the World* reports, both in hardback and paper, since 1984. These information-rich books, translated into nine languages and widely disseminated, interpret their findings against the yardstick of "sustainability." In addition to these books, the institute has a *Worldwatch Paper* series with over ninety titles in all. (Annual subscriptions to this series cost twenty-five dollars; a current *State of the World* book is included in the subscription cost.) Topics covered in the first sixteen years include energy efficiency, population growth and migration, food, changing roles of women, recycling, global economy, impact of technology, soil erosion, nuclear power, toxic wastes, and deforestation. *Worldwatch Magazine,* inaugurated in 1988, is published six times a year and is available, by subscription, for twenty dollars. The goal as stated in the November–December 1989 issue is to "reverse the environmental trends that are undermining the human prospect."

The World Resources Institute, founded in 1982 with Gus Speth as the president, is a policy research center to help governments, international organizations, and the private sector address vital issues of environmental integrity, natural resource management, economic growth, and international security. The four program areas of the institute include (1) climate, energy, and pollution; (2) forests, sustainable agriculture, and biodiversity; (3) economics and institutions; and (4) resource and environmental information. A new Center for International Development and Environment provides policy advice and technical assistance to local groups charged with managing natural resources and economic development. The *World Resources* report, published since 1986, is an authoritative reference book for global statistics (data tables make up one-quarter of the book), even though the data is somewhat oriented to the United States. Presumably, this slant is unavoidable, as data is more available in the United States than in other countries. But the reports are more than just facts and figures; people and social issues are the focus throughout, and there is good use of case study material. Cross-referencing between chapters helps to highlight the important linkages among the various aspects of the environment, and the references in each chapter provide an excellent entry into current literature.

—BARBARA K. RODES

"Groups organize most readily to defend their resource base against the incursion of others, but they may also organize to reverse deterioration driven by forces internal to the community. As Kenya's forests shrink, for instance, thousands of women's groups, youth clubs, and harambee societies have mounted local tree planting drives. The National Council of Women of Kenya inaugurated its Greenbelt Movement in 1977, calling on women's groups across the country to turn open spaces, school grounds, and roadsides into forests. Over a million trees in 1,000 greenbelts are now straining skyward, 20,000 mini-greenbelts have taken root, and upwards of 670 community tree nurseries are in place. Meanwhile, Kenya's largest women's development network, Maendeleo Ya Wanawake, with its 10,000 member groups, initiated a campaign in 1985 to construct wood-saving improved cookstoves."

—State of the World 1989

"*Uganda Installs a Geographical Information System*

"Uganda, seeking to improve its collection and management of information on natural resources, completed the pilot phase of a new Geographical Information System (GIS) in 1987. Although basic data on natural resources had been collected previously, much of it was scattered, stored in incompatible formats, or quietly gathered dust, unknown to public and private resource managers.

"GIS technology promises to solve some of these problems, and thus increase the usefulness of natural resource data. GISs are a family of specialized computer software that allow geographically-oriented data (such as political boundaries, forest locations, or population centers) to be housed together on a set of computerized maps. GISs represent a quantum leap in cost-effective data management and analysis; they combine data from diverse sources and can analyze and display the relationships among many environmental parameters. (For additional information, see *World Resources 1987*, Chapter 11, "Global Systems and Cycles.")"

—World Resources 1988–89

Encounters with the Archdruid
JOHN MCPHEE
1971, Farrar, Straus and Giroux

John McPhee's *Encounters with the Archdruid* is far more than clever, but you have to admit that getting archdruid David Brower out onto three conservation battlegrounds, with as many opposing generals, was pretty shrewd biography.

The encounters between Brower, now chairman of the Earth Island Institute (and indisputably America's greatest living conservationist) and his three adversaries—Commissioner Floyd Dominy, archbeaver of the Bureau of Reclamation; Gordon Park, genius geologist and mining-company director; and Charles Fraser, "enlightened" developer—are more than rhetorical clashes. They take place on the Colorado, in the North Cascades, and on Cumberland Island, and they not only reveal the psychology of the combatants and describe the real stakes—unspoiled landscapes—in those stereotypical conflicts over dam building, wilderness mining, and coastal development, but do a pretty good job of suggesting the extraordinary character of David Brower.

McPhee's book was written in 1971, shortly after Brower's ouster from the helm of the Sierra Club and his founding of Friends of the Earth. Brower's luster was perhaps then its brightest. Despite the subsequent lull in the popularity of conservation, ecology, and environmentalism, however, Brower has continued in the vanguard, kicking up clouds of organizational, as well as ideological, controversy.

Through his leadership in the Fate of the Earth conferences, Brower has helped to link ecology and disarmament. And lately septuagenarian Brower was photographed planting a tree in Nicaragua, where the most recent Fate of the Earth conference was held. Its focus was on restoration, peace and global security, and human rights, issues relevant to the nonindustrial South. Showing up in Central America to address such concerns is not all that typical of conservation executives, but enlarging the compass—and meaning—of the movement is what Dave Brower has been doing from day one.

—STEPHANIE MILLS

" 'The future can take care of itself,' Park said. 'I don't condone waste, but I am not willing to penalize present people. I say they're penalized if they don't have enough copper. Dave says they're penalized if they don't have enough wilderness. Right?' He smacked a stone with his pick."

"Brower was feeling good, too—obviously enjoying himself on the island. Why he did not rise up and clout Fraser, verbally, seemed a little odd to me, but I had seen him before in situations where he was getting the sense and feel of something, and while his mind was working toward a settled attitude he had vacillated or lapsed into an uncharacteristic passivity. In the North Cascades, he had known where he was. He had been there before, and had fought for the wilderness there. He had never before set foot on Cumberland Island."

" 'Reclamation is the father of putting water to work for man irrigation, hydropower, flood control, recreation,' Dominy said as he turned on the lights. 'Let's *use* our environment. Nature changes the environment every day of our lives—why shouldn't *we* change it?' "

" 'Commissioner,' I said, 'if David Brower gets into a rubber raft going down the Colorado River, will you get in it, too?'
" 'Hell, yes,' he said. 'Hell, yes.'

    *   *   *

"Mile 130. The water is smooth here, and will be smooth for three hundred yards, and then we are going through another rapid. The temperature is a little over ninety, and the air is so dry that the rapid will feel good. Dominy and Brower are drinking beer. They have settled into a kind of routine: once a day they tear each other in half and the rest of the time they are pals."

EDWARD ABBEY

I was the only one out of my circle of friends at college to go to war. I was also the only one of my peers who had done any camping and gone on primitive walkabouts, which were later known as backpacking. When I got back from Vietnam in 1968, all this had changed: My ex-roommate, who had never camped out in his life, had become a big Sierra Club guy and serious hiker and climber. He spent hours explaining Vibram soles, stuff bags, and other innovations in the booming and brand-new outdoor equipment industry. I received his indulgence and realized I was now behind the times in the only thing I had ever been ahead in. The reason for all this was Edward Abbey. He had written a book called *Desert Solitaire*. This book changed lives.

I wince as I type the name of the man I helped bury this spring. But even if I had not been close to him for nineteen years, it would be hard to objectively measure the impact of Edward Abbey's art and politics. Ed Abbey published seven novels, seven volumes of essays, and several coffee-table collaborations before he died in March of 1989. He also had roughed out a posthumous novel, *Hayduke Lives*, a sequel to *The Monkey Wrench Gang*, the fine comic novel considered by some the most influential—certainly the most notorious— book among those shaping the modern environmental movement. His death, as was much of the significance of his work throughout his life, was largely ignored in the East. The funeral notice in *Backpacker* claimed he preached to the already converted. But that is quite incorrect: Without Ed Abbey, *Backpacker* magazine, along with many of the underlying attitudes and interests which looked to the wilderness, would perhaps not exist. The man so stirred people sometimes without their knowing. His presence was that important.

Abbey's goals were simple: save the world but not allow it to become more than a part-time job. He decided to become a writer. He saw himself primarily as a novelist and as early as 1956 staked out the literary territory he never would leave, with the publication of one of his best novels, *The Brave Cowboy*, a story of tyranny, freedom, and resistance. His preference for regarding himself as a writer of fiction accounts for irritability toward the disproportional success of *Desert Solitaire*, his first nonfiction book.

Ed was variously called the Thoreau of the American West, one of a kind, an American artifact, a man with the bark still on, a clear-eyed wilderness addict, and a happily excessive, iconoclastic refugee from overcivilization. His work was labeled enraged, hilarious, vulgar, passionate, romantic, ornery, xenophobic, arrogant, elitist, misanthropic and sentimental, entertaining, and exasperating. There is truth in all this.

Ed Abbey was neither a naturalist nor an environmentalist,

though he writes of and defends some of the same things. *Desert Solitaire* was something larger than just a book about the desert. It was about the power of the land, the human connectedness to the Earth, an idea of freedom—it was a call to arms. This unplanned book, with minor exceptions such as the three pages on music, simply slipped out of the right side of his brain and seemingly wrote itself. Ed Abbey could never quite accept it for what it was—an American classic.

Abbey was a writer who wrote of the death of America while fighting for its life. He was a writer like Celine, whom he admired. His work, especially the essays, was autobiographic and spoke of "the writer's duty to hate injustice, to defy the powerful, and to speak for the voiceless." His art and his politics were inseparable. In twenty years, his voice never tempered.

You must read the books. *Desert Solitaire,* of course. There is an equivalent masterpiece among his essays, such as "Down the River" or "Beyond the Wall." The novels are all fun. It's difficult for me to believe he wrote *The Brave Cowboy* during the days of Eisenhower and pinko Commies. Since that time (1956), in fact, I can't recall a single dull piece, and believe me I've read them all.

Like other excellent writers, he was a prodigious walker. Abbey traveled less widely than some, but he saw clearly and wrote with more fortitude and honesty than all but a handful of his contemporaries of the suffering and destruction seen everywhere on Earth. He spent more time sleeping under the stars than any living writer perhaps excepting a Peter Matthiessen in his earlier years.

What set Abbey apart—even from writers of distinction like John McPhee and Barry Lopez—was that he saw the absolute danger of our times, the global crisis, the planet burning from a way of living encouraged and praised by governments and corporations. The world is on fire and Edward Abbey was here—is still here in his books—to get angry and make us mad too. He told us the truth and what values we might hang on to save ourselves and this Earth we want to abandon for space and "yearn for some more perfect world beyond the sky. We are none of us good enough for the world we have."

—DOUG PEACOCK

57

Desert Solitaire: A Season in the Wilderness
1968, Simon & Schuster

The Monkey Wrench Gang
1975, Avon Books

The Brave Cowboy
1956; 1977, University of New Mexico Press

" 'This would be a good country,' a tourist says to me, 'if only you had some water.'

"He's from Cleveland, Ohio.

" 'If we had water here,' I reply, 'this country would not be what it is. It would be like Ohio, wet and humid and hydrological, all covered with cabbage farms and golf courses. Instead of this lovely barren desert we would have only another blooming garden state, like New Jersey. You see what I mean?'

" 'If you had more water more people could live here.'

" 'Yes sir. And where then would people go when they wanted to see something besides people?' "

—Desert Solitaire

"Hayduke jumped off. As the tipover point approached the tractor attempted (so it seemed) to save itself: one tread being more advanced into the air than the other, the tractor made a lurching half-turn to the right, trying to cling to the rim of the mesa and somehow regain solid footing. Useless: there was no remedy; the bulldozer went over, making one somersault, and fell, at minimal trajectory, toward the flat hard metallic-lustered face of the reservoir. As it fell the tracks kept turning, and the engine howled.

"Hayduke crawled to the edge in time to see, first, the blurred form of the loading machine sinking into the depths and, second, a few details of the tractor as it crashed into the lake."

—The Monkey Wrench Gang

AIR

T AKE A DEEP BREATH. Go ahead. Inhale till you feel your diaphragm move. What do you feel? Is it the blessed breath of life? *Prana* is what the yogis call it—breath energy. The act of breathing spans the conscious and unconscious, voluntary and involuntary body functions. Hence, focus on the breath is an ageless meditative practice. Rightly understood, every breath you draw is a gift from the universe. We are air-breathing animals. To live we need oxygen. In just the right amount Earth's atmosphere provides the oxygen molecules to help us metabolize our sustenance and the inspiration that animates us.

The air seems almost like nothing. It is the membrane through which we perceive, and by which we are protected from, the sparkling vacuum of the cosmos. The air is the enveloping atmosphere of planet Earth, inviting our sight to seek the heavens. Sometimes (or seemingly always, depending on where you live) the air bears countless tons of water, as cloud panoramas or as just an expanse of overcast (when as far as one can see is up to the ceiling). Thanks to the jet age, many of us have had the experience of looking at clouds from both sides now. We've passed through them and flown over them and cringed as pilots threaded their ways among them. Many more of us have watched the skies, with the realms of clouds, just for pleasure. Down to Earth, on a bright day, you may look out to a clear blue vastness, all illumined by the sun, simile of an empty mind. And at night, if you are fortunate enough to live in a region where ambient light doesn't interfere, you can gaze up at the stars, knowing that there are billions of them, and

61

possibly millions of solar systems, some that might even be hospitable to life.

Because we humans are always looking for guidance (or justification), usually in the form of a story, it is perennially human to try to read the metaphors in the night sky, to tell the myths associated with the constellations, and to seek the governance of the zodiac, or of fateful shooting stars, or comets as portents of millennial change.

Migratory birds also consult the sky for guidance, if not of the metaphysical sort. They navigate by star patterns and the position of the sun on their journeys the length of a hemisphere, or across trackless seas. Before we ventured up in balloons or 747s, the sky was largely reserved to the stunning variety of species of birds. However common, a bird is always a bit of a wonder. Even the neighborhood birds are a delight and an amusement. One late autumn afternoon in the course of writing this essay, I watched an impromptu convention of about a dozen crows taking place in a taller-than-average tree out back in the lately bare woods. These crows were a lively crowd; they seemed a little precariously perched, bobbing in the wind as the branches swayed in the chilly breeze. I love the way crows seem to drift, float, and quiver in midair. A couple of them will slope off, casually doing a loop-de-loop together for no other purpose than fun and sociability; then others lift off and alight, taking a quick look-see around the treetops. The why of it is a happy mystery.

As the province of flying things, the air was, until only very recently, a realm beyond our powers. Now *we* claim to have conquered the sky—flight paths have become to us what sea-lanes were to our great-grandparents. Not only have airplanes become a commonplace means of transportation in the late twentieth century, in space capsules we have escaped Earth's ambit and traveled far enough away to have a look back at the home planet, and take snapshots of Earth as Gaian mandala.

Earth's enveloping atmosphere is the caldron of planet-girdling currents of wind, humidity, and temperature, the arena of tornadoes, blizzards, and drought. The weather is a mix of water and air driven by fire (in the form of solar radiation). Impeded and rerouted by Earth's landforms, given its dynamic by temperature differentials and the Earth's rotation, weather governs our lives. Over evolutionary time each living thing has developed a specific relationship to the weather and climate. Most organisms are not so cosmopolitan as we are. Plants, being unable to flee, are particularly limited in their temperature ranges. Animals, being more mobile, are less limited by climate, except those creatures that are closely coevolved with a single plant species or associations. And *Homo sapiens,* who can build shelter and make clothing, has extended her range throughout the planet. Because of the rootedness of plants, dramatic or sudden changes in the weather can be devastating, ecologically, to wild plant communities. Global warming, brought about by a man-made increase in the atmosphere's "greenhouse" gases, is exactly such a change. Cosmopolitan and footloose though we may be, because our civilization depends on agriculture, and a relative handful of cultivated plants, also entirely subject to climate, variations in weather hold the potential to change history.

Weather is made of many forces. Among these forces humans have come to know the winds as intimately as anything unpredictable and invisible can be known. We have been influenced by the winds, have blessed and cursed them: hot winds that make people do crazy things, winds that create conditions for wildfires or set the undertone of certain seasons, have names—chinook, simoom, samiel; foehn; khamsin, harmattan, sirocco, solano, Santa Ana—native to the places where they blow.

Some winds change the face of the Earth, working over the eons, subtly blasting grains of sand against rock faces, sculpting out magical shapes—pinnacles, needles, and arches. Winds move huge sand dunes, tons of mass grain by grain, in sinuous ranks. The magnitude of the wind as an Earth-shaping force is plain in phenomena like windblown soils in Tennessee that were first ground to powder by glaciers in China.

In addition to flecks of inorganic matter, winds carry other small particles—pollen and spores—moving germs of life around, colonizing new places, and abetting the great cause of genetic variation. In the back eighty I can watch this transport as autumn gusts gather up seedheads of switchgrass and tumble them in the air. The wind delicately lifts them high overhead, then piles them in blond thatchy drifts against wire fences and borders of close-set firs. It carries wisps of milkweed floss, causes leaves to tremble, rips them free, whirls them up in little cyclones, and, after a short dance, abandons them to the Earth to settle and begin more soil.

Bob Dylan, the great bard of the sixties generation, sang that you don't need a weatherman to know which way the wind blows. True enough. If wind there be, you can just step right out and feel it on your face. Your breeze is a local phenomenon, local as the canyon between the high-rises, or the little hill you're perched upon, local as your cheeks and their nerve endings, as the fine hairs clothing your skin. The wind that spawns the breeze may come from half a world away, however. We live in an era where technology lets us *see*, from twenty-two thousand miles up, which way the wind blows across the hemisphere and around the planet. Interpreting and purveying that consciousness-altering imagery is the happy task of many a television weatherperson.

Dylan notwithstanding, who better to consult about air than a weatherman? (There just happened to be one—articulate, approachable, concerned—right down the road.) So one morning just before Halloween, on a glorious balmy hazy day (one of the Michigan weather's sucker punches), I headed east down Highway 72 to station WPBN, near Traverse City, for a talk with Dave Barrons, who does the noon and six o'clock weather reports there. Not just another talking head, Barrons has lately immersed himself in a study of global warming, and is actively seeking local audiences to address on the subject of this epochal man-made change in Earth's atmosphere and climate.

During our visit, Barrons took me into his corner of the newsroom, where there's a TV that displays, in lividly enhanced color, the satellite weather map and a VDT scanning line after line of space-time coordinates cataloguing, over several hours, the series of infrared images provided by the satellite. He showed me how he programs his computer to animate the series of images, which that day resulted in a time-lapse movie of cloud masses rolling northeastward, over the central United States, to be thwarted by a large mass of high pressure. "This is a high-tech science that the public gets to savor every night," he said.

Although satellite imagery has provided us with the big picture of weather systems, it hasn't entirely supplanted land-based observers who check rain gauges and anemometers and phone the results in to a network. Other people whose lives depend, one way and another, on the dynamics of the weather—farmers and sailors—are also keen readers of the signs written in the shapes of the clouds, the hue of the dawn, the shifting of the winds. And many of us, with no training at all, can anticipate some barometric pressure changes simply by feeling them in our bones.

However much we might seek to, there's no controlling the wind. It bloweth where it listeth, which is all the more reason that we should refrain from burdening it with pollution. Airborne toxics (volatile organic chemicals, petrochemicals, and solvents, mainly) have become, strangely enough, a major source of water pollution. (According to one authority, half the pollution in the Great Lakes comes from northward-drifting pesticides sprayed in Mississippi and Louisiana.) These ubiquitous and largely unregulated poisons range from wafted hydrocarbons—PCBs and toxaphenes in fish far from any detectable point source of contamination—to lead and particles of asbestos now found in human tissues as well as in the water supply. Once in the wind toxic threats become diffuse, but no less certain. It does become harder to fix the blame and levy the fine when any commons—land, water, or air—is abused.

The threats to our common atmosphere from civilization's industrial activity and the burgeoning of human population are undoubtedly the most staggering of all the ecological crises we face: ozone depletion, acid rain, and global warming. They may soon render the atmosphere and biosphere hostile to the continuation of the kind of earthly life we enjoy. In addition to changes in the composition of the living atmosphere, there are also specific physical and chemical threats posed by other airborne contaminants. If the wind can carry seeds and spores and the fragrance of hay, it can also carry man-made molecules, the gases and compounds exhaled by industrial activities. We have, in fact, depended on it to do so, building higher smokestacks in hopes that the air currents farther up will carry our smokes farther away, out of sight, out of mind, eliminating them as a local problem, at least. However, because the atmosphere is essentially a closed system, displacement and dispersal are not really disposal. Now we are learning that some of the chemicals contaminating the air, however invisible or barely perceptible, are doing immense damage.

It almost makes one nostalgic for smog, good old-fashioned air pollution as we knew it back around the original Earth Day. Smog we could perceive directly through our senses. (The installation of catalytic converters in auto exhaust systems has helped to reduce the per auto emissions of substances that make smog. However, the sevenfold increase in the numbers of automobiles around the world since the early fifties has overwhelmed even those improvements.) Now the most serious atmospheric pollution is less immediately obvious and far more insidious than urban or industrial smog.

Vying with global warming is the threat of atmospheric decomposition. We face rapid and dramatic changes in the makeup of the stratosphere as a result of ozone depletion, which permits an increased amount of ultraviolet radiation to reach the Earth's surface. When the original Clean Air Act was drafted, CO_2 emissions weren't dealt with, partly because the global-warming consequences of increasing the CO_2 content of the atmosphere were, at the time, still speculative, but mainly because CO_2 release is so fundamentally a part of energy use that the only way to reduce CO_2 release is through dramatic gains in combustion efficiency. We must conjure with the greenhouse effect—an increase in the amount of atmospheric carbon dioxide due to combustion, and in the amount of methane due to food production, especially cattle raising. We also have produced changes in the pH of the hydrosphere, a consequence of acid deposition.

Tall stacks on coal-fired electrical generation plants and on metal smelters blast compounds high into the atmosphere that when cooked by sunlight and mixed in water droplets make sulfuric and nitric acid, which falls to Earth in the form of acid snow, acid fog, or acid rain. The whole syndrome is summed up as acid deposition. Acid deposition has devastated forests in central Europe, and mountain groves in northern New England and Scandinavia. A subtler but no less ecologically damaging effect of acid rain is the clear sterility of lakes whose aquatic fauna has been wiped out by pH changes. The puzzling occurrence of dead lakes in New York's Adirondacks and throughout Sweden and Norway led to some of the first research into the sources and consequences of acid deposition. Just as most creatures are limited to a certain climatic range, so most aquatic life is bound within a certain pH—p(otential of) H(ydrogen), a measure of relative acidity or alkalinity. Thus fallout from our wasteful patterns of energy and resource consumption, including the electric-power industry's resistance to adopting cleaner, more efficient processes, is making life impossible for the bright flashing trout and other inhabitants of lakes in affected regions.

Another form of pollution that is leading to actual atmospheric destruction is caused by the release of CFCs, or chlorofluorocarbons (and, to a lesser extent, halons, which are used in fire extinguishers). CFCs, first hailed as wonder chemicals for their lack of toxicity, are widely used as blowing agents in rigid plastic foams, as refrigerants, and as propellants in aerosol dispensers. As the foams decompose and the Freon escapes from the air conditioners, CFCs, over a period of decades, make their way into the upper atmosphere, where their chlorine atoms cause molecules of ozone to decompose. This depletion of stratospheric ozone is exposing the biosphere to an increased amount of ultraviolet radiation. The ozone layer, now being compromised, helps to maintain conditions favorable to life on Earth. Solar radiation fosters life, but too much solar radiation from the ultraviolet end of the spectrum can cause eye cataracts and cancer, damage crops, and depress immune systems. That this is taking place, and that CFCs and halons are causing it, is not a matter of speculation.

The global community of scientists has urged a speedy end to the production and release of CFCs. Even if this were to be achieved today (which seems highly unlikely given foot-dragging on the part of the many industries manufacturing and utilizing these chemicals, and to the pressure they exert on their governments), CFCs and halons now in use will find their way upward and will cause still further destruction of the ozone layer. Reality is that the ozone depletion measurable *now* is the result of emissions from thirty years ago.

How may we hope to restore the air's quality? Regulation of industrial discharges into the atmosphere has helped. Further changes in industrial practice are possible. And nation-states can work together more effectively to protect our shared atmosphere.

Existing clean air legislation in the United States dates from 1967. Our knowledge of the problems has changed dramatically since then. Much has been learned since then about the components of air pollution/atmospheric pollution and ways and means of reducing them. The original Clean Air Act was written before we understood the vast consequences of changing the mix of gases in the atmosphere. It sought to reduce emissions of suspended particulate matter, sulfur oxides, carbon monoxide, nitrogen oxides, ozone, volatile organic compounds, and lead—and established limits for stationary sources of air pollution.

Over the intervening decades, a whole new generation of problems has come to our attention, and has yet to be successfully addressed by Congress. The auto industry, whose influence in Congress is formidable, managed to block reauthorization of the Clean Air Act for years to impede adoption of stricter emissions standards. As of this writing, that deadlock has finally been broken.

As a long-term strategy for cleaning the air, regulation is a little like closing the barn door after the horse is lost. David Stead, of the Michigan Environmental Council, echoed the thoughts of many on the value of environmental regulation per se: "All it really does is create a holding action. . . ." It is crucial, but ultimately palliative. It may buy time to approach the far thornier question of how not to produce and spread toxics into the atmosphere (and hydrosphere) in the first place.

Human creativity, strong community organizing, and late-breaking altruism can come into play to answer that question, with actions ranging from consumer abstinence and awareness to outright legislative bans. But the urgency of taking action at all levels, from individual households to the international treaty tables, must first be widely understood and felt, sharply, as a caustic vapor in the lungs (and by the committed and formidable citizen organizers of the toxics movement, it has). In protecting this element, as all others, the hope is that we will place the preservation of the very air, which of all Gaia's systems unites us most intimately, over our all-too-human reluctance to change our lifestyles and the industrial civilization that undergirds them. These are not easy, or trivial, choices. Our continuation as a species, however, and the continuation of the life of our home planet, depend on our choosing rightly, and soon.

Apropos such change, says Jessica Tuchman Mathews, vice president of the Washington-based World Resources Institute, "The one thing we can be absolutely sure of is that the sooner we act, the less costly, the less painful it will be."

The slight glimmer of good news in all of this is that if we halt the destruction of the ozone layer and the increase in atmospheric CO_2, in time the Gaian processes that originally created an atmosphere so right for us and our companion life-forms may be able to replenish it.

The Gaia Hypothesis may become the basis for a revolution in our understanding of Earth, and of Nature. The scientific insight that led to this formulation came from noticing that the chemistry of Earth's atmosphere, relative to that of other planets in the solar system, is anomalous. Based on the reactivity of its gases, Earth's atmosphere is not what science would predict. This is the most telling clue that Earth functions like an ecosystem, if not an organism. Chemist James Lovelock began to ponder these anomalies and eventually, in collaboration with microbiologist Lynn Margulis, elucidated and proposed, circa 1975, the Gaia Hypothesis.

The Gaia Hypothesis asserts that Earth's atmosphere is continuously interacting with geology (the lithosphere), Earth's cycling waters (the hydrosphere), and everything that lives (the biosphere). Evidently the atmosphere has always been an integral factor of the evolution of life. It seems not to have been passive, but an active chemical medium participating in evolution. The atmosphere has been changed dramatically by the appearance of certain life-forms (the cyanobacteria, for instance), which, incidentally, created conditions more favorable to other, subsequent life-forms—like us.

Thus the hope is that the living air itself will offer some forgiveness: Gaia's autopoiesis. (Autopoiesis means self-making—"The concept is that it is intrinsic in cells and organisms to maintain their organization via interactions with their environment," explains Gaia Hypothesis coauthor Lynn Margulis.) The image is that the atmosphere is a circulatory system for life's biochemical interplay. If the atmosphere is part of a larger whole that has some of the qualities of an organism, one of those qualities we now must pray for is resilience.

A Field Guide to the Atmosphere
(The Peterson Field Guides: number 26)
VINCENT J. SCHAEFER AND JOHN A. DAY
1981; 1983, Houghton Mifflin Company

As with all the Peterson Guides, once published, *A Field Guide to the Atmosphere* instantly became a standard reference.

It won't instruct the amateur weather watcher about such current issues as global warming, but the *Field Guide to the Atmosphere* does provide all of the basics for understanding the chemical properties and physical processes of the air around us. (It even includes suggestions for experiments, hence makes an excellent self-study course.)

Especially good are the photographs and descriptions of clouds, indispensable for learning how to read the sky. Clouds are not only the most visible feature of the atmosphere, they are often the most spectacular. Dr. John Day's 358 photographs, 32 in color, of every kind of cloud type are superb.

I wrote this review while watching a beautiful rosy sunrise on a winter morning. From the point where the sun would break the horizon, a vertical column of color, called a sun pillar, rose high into the sky. The Peterson Guide explained that this glory is caused by flat hexagonal ice crystals falling with "their larger axis horizontal, each crystal acting like a tiny mirror."

From airborne particles to temperature lapse rates to severe storms, the very small and the very large atmospheric phenomena are all discussed succinctly. It's not quite bedtime reading (for one thing, the print is very small), but if one has room, or time, for only one atmospheric handbook, this would be the single best.

—DAVE BARRONS

"Firestorm

"One of the most awesome of natural atmospheric phenomena, similar in many respects to the tornado, is the firestorm. This often develops when a forest wildfire becomes organized by a cyclonic circulation. The highly unstable air (greatly exceeding the dry adiabatic rate) pulls in the surrounding air and develops a massive fire of terrifying proportions. An organized firestorm creates its own localized wind pattern which often becomes so violent as to fell trees, tear burning limbs from them, and scatter embers from the upper levels of the convective column far and wide, thus starting new fires."

The Home Planet
EDITED BY KEVIN W. KELLEY
1988, Addison-Wesley

James Lovelock, author of the Gaia Hypothesis, has said that the most outstanding outcome of the NASA space program is not the overhyped technological spin-offs (perfect ball bearings and non-stick frying pans), but the chance to see the Earth from space for the first time in human history. And as we page through this book, we get a growing sense why this might be so. One hundred and fifty stunningly beautiful color photographs of Earth culled from the best available from all U.S. and U.S.S.R. space missions are lovingly interlaced with excerpts from the words of their astronauts. These excerpts range from mundane observation, to the touchingly human anecdote, to attempts to convey in words the experience of full-blown cosmic revelation. In these lines (and very strongly in between them), we start to get a sense of what it is about the distant glimpse of earthlight which so consistently transforms hard-boiled technicians, engineers, and physicists into dewy-eyed metaphysicians and poets. Why, upon their return, they feel themselves to be the miraculously privileged sensing elements of the species, duty-bound to convey something to the rest of us, something so important that it almost can't be talked about yet must be.

—YAAKOV GARB

"They say if you have experiments to run, stay away from the window. For me, preoccupied with the Stop Dynamics Module, it wasn't until the last day of our flight that I even had a chance to look out. But when I did, I was truly overwhelmed. A Chinese tale tells of some men sent to harm a young girl who, upon seeing her beauty, became her protectors rather than her violators. That's how I felt seeing the Earth for the first time. 'I could not help but love and cherish her.'"

—TAYLOR WANG
China/United States

"One morning I woke up and decided to look out the window, to see where we were. We were flying over America and suddenly I saw snow, the first snow we ever saw from orbit. Light and powdery, it blended with the contours of the land, with the veins of the rivers. I thought—autumn, snow—people are busy getting ready for winter. A few minutes later we were flying over the Atlantic, then Europe, and then Russia. I have never visited America, but I imagined that the arrival of autumn and winter is the same there as in other places, and the process of getting ready for them is the same. And then it struck me that we are all children of our Earth. It does not matter what country you look at. We are all Earth's children, and we should treat her as our Mother."

—ALEKSANDR ALEKSANDROV
Soviet Union

"For the first time in my life I saw the horizon as a curved line. It was accentuated by a thin seam of dark blue light—our atmosphere. Obviously this was not the ocean of air I had been told it was so many times in my life. I was terrified by its fragile appearance."

—ULF MERBOLD
Federal Republic of Germany

Gaia: A New Look at Life on Earth
JAMES LOVELOCK
1979; 1987, Oxford University Press

The Ages of Gaia: A Biography of Our Living Planet
JAMES LOVELOCK
1988, W. W. Norton and Company

We don't view a cat's fur or skin or blood as some haphazard cluster of nonliving excreta which just happen to cloak that part of the creature which is "truly" living; why then should we regard the Earth's atmosphere—which is produced by life on Earth and serves to protect it and circulate its vital elements—as random and inert? Wouldn't it be more accurate to speak of the whole planet, including its atmosphere, soils, and oceans, as a single living organism?

This was the conclusion of scientist James Lovelock, who was hired to work on instrumentation for the detection of life on Mars for NASA in the 1960s. Backtracking, he asked a more fundamental question: Assuming that life elsewhere won't necessarily be similar to the forms we're familiar with on Earth, how will we recognize life per se when we see it? Using a working definition of life-as-entropy-reducing-process, he next realized that the atmosphere was both a product of—and protector of—life. (For instance, while the sun's output has increased over the last several billion years, the atmosphere's composition has adjusted in order to remain a protective shield for life on the planet.) Known as the Gaia Hypothesis, Lovelock's view of the Earth as one organism has been annoying and inspiring scientists for almost two decades now.

Lovelock's first book, *Gaia: A New Look at Life on Earth,* begins with a review of the formation of the planet, then describes the central evidence for the Gaia Hypothesis, some of the cybernetic ideas in which it's grounded, and some of the key processes in Gaia's functioning. It ends with some reflections on our place within Gaia, for if we are to really accept Lovelock's ideas, we should think of ourselves as living *in* rather than *on* Earth.

His recent *The Ages of Gaia: A Biography of Our Living Planet* is a more mature book, which incorporates subsequent work by Lovelock and others on the ramification of the Gaia Hypothesis, his response to critics, and more recent information about the environmental crisis.

The Gaia Hypothesis is a whole range of things at once, hence the cloud of acrimony and enthusiam which doggedly accompanies it. On the one hand, it's a serious and testable hypothesis in planetary astronomy; on the other hand (and this has not made the winning of scientific credibility any easier), it's a nucleus thrown into a culture desperate for alternative organizing metaphors through which to understand its relationship to nature. It is not so much the proposition of Gaia itself which has caused such a stir, it seems, but the way in which the mind uses it to leap out to other things. Read these books and encounter Gaia for yourself.

—YAAKOV GARB

"When methane reaches the atmosphere it appears to act as a two-way regulator of oxygen, capable of taking at one level and putting a little back at another. Some of it travels to the stratosphere before oxidizing to carbon dioxide and water vapour, thus becoming the principal source of water vapour in the upper air. The water ultimately dissociates to oxygen and hydrogen. Oxygen descends and hydrogen escapes into space. By this means a small but possibly significant addition of oxygen to the air is ensured in the long term.

When the books are balanced, an escape of hydrogen always means a net gain of oxygen."

—*Gaia*

The Coevolution of Climate and Life
STEPHEN H. SCHNEIDER AND RANDI LONDER
1984, Sierra Club Books

Since the publication of *The Coevolution of Climate and Life* in 1984, Dr. Stephen H. Schneider has become a recognized authority on the controversial and topical issue of global warming. Senator Tim Wirth has called him "the world's leading expert" in that field. He is a frequent advisor to Congress and in 1989 published a book entitled *Global Warming: Are We Entering the Greenhouse Century?* This later book is considerably shorter and less comprehensive than the earlier work, however, and should not be considered to supersede it.

The first half of *The Coevolution of Climate and Life* consists of an overview of the Earth's four billion years' climatological history, entitled "Climate Before Man" [*sic*]. This is followed by an extensive description of the global climate system, how it works, and the causes of change. The second half of the book is devoted to the human impact on climate and the resultant effects this will have on food, water, health, and overall human welfare.

The Coevolution of Climate and Life remains a valuable and accessible text for those of us wanting both to understand better the climatological workings of the world and the dangers inherent in ignoring the effects unchecked human activity will have on the continued "coevolution of climate and life."

—NANCY JACK TODD

"But what has societal complexity and its relationship to democracy to do with the weather? We hope we've convinced you that, at least indirectly, the weather has quite a bit to do with societal vulnerability to shortfalls in food, water, and energy supplies. These vulnerabilities, in turn, are related to complexity and maintaining a vital democracy. The climatic connection becomes clearer when—because of special interests' pressure or average citizens' indifference or ignorance—societies permit food security margins to become so small that price instability and even famines can result from precedented climate fluctuations. Moreover, classic, market-oriented economic practices that do not internalize externalities or protect common property can lead to the erosion of soil, exclusion of pollution-caused health problems from the cost of doing business, or underreaction to the possible consequences of long-term climatic changes from worldwide industrial and land-use activities. And because these issues are complex, we must make a considerable effort to understand related probabilities and consequences. From such knowledge we can then choose through the democratic process how to allocate resources to mitigate climatic hazards or to achieve any other societal goal. Society largely determines the societal consequences of weather variations."

"Among the nastiest surprises a hurricane holds for people is the *storm surge*. Since pressure at a hurricane center falls below that of the surrounding area, it raises sea level by sucking up the water into the storm's center, much like a gigantic straw. This effect, combined with winds that can reach 320 kilometers (200 miles) per hour, can temporarily raise sea levels by several meters, inundating coastal areas as the storm hits the coast. The catastrophic flooding that results typically takes more lives and ruins more property than any other aspect of the storm."

Global Warming: Are We Entering the Greenhouse Century?
STEPHEN H. SCHNEIDER
1989, Sierra Club Books

The Challenge of Global Warming
EDITED BY DEAN E. ABRAHAMSON
1989, Island Press

Two new books offer easy access to the difficult subject of global warming: Dr. Stephen Schneider's *Global Warming: Are We Entering the Greenhouse Century?* and *The Challenge of Global Warming*, edited by Dean Edwin Abrahamson. The answer of both books to the question posed by Schneider's title is a resounding yes. "We have become a competitor with the natural forces that drive our climate," says Schneider. The changes we are causing are becoming both irreversible and so rapid that many natural and societal systems may not be able to handle the rate of change.

In *Global Warming* Schneider elaborates the complexities of climatological theory—including the considerable scientific uncertainties—with remarkable lucidity and balance. The book opens with an excellent chapter projecting what life in the United States might be like partway into the "greenhouse century" and then hopscotches between scientific explanation and interesting behind-the-scenes glimpses of the international scientists, politicians, and media personalities that help make this issue so controversial.

The Challenge of Global Warming is a collection of some of the more important papers, articles, and conference statements on the subject. It is an ideal resource on global warming for public officials, policy makers, students, environmental activists, and concerned citizens. It offers probably the best in-depth overview of the subject currently available and includes work by James Hansen, George Woodwell, Roger Revelle, and Rafe Pomerance, as well as other

environmental scientists and policy experts. The introduction is by
Senator Timothy Wirth.

—BILL PRESCOTT

"How important is a degree of temperature change? . . . A degree
or two temperature change is not a trivial number in global terms
and it usually takes nature hundreds of thousands of years to bring it
about on her own. We may be doing that in decades. . . . Humans are
putting pollutants into the atmosphere at such a rate that we could
be changing the climate on a sustained basis some ten to hundred
times faster than nature has since the height of the last ice age."

"The prospect of whole communities disappearing in the U.S. high
plains on top of the rapidly diminishing groundwaters of the
Ogallala aquifer in eastern Colorado, western Kansas, the Texas
panhandle, and northeast New Mexico raises major social and economic questions for the future."

"There is a tension between the scientific culture of caution and
reticence and the media's penchant for drama, dread, and debate
that keeps the show lively and the audience tuning in . . . there is a
growing mismatch between the complex nature of reality and the
way such problems are usually reported in the popular media or
perceived by the public."

"How can the present market, which determines the price of
electricity, coal, and so on, possibly represent the true cost of these
fuels or the energy produced from them if there is no charge for the
environmental damage they do?"

—*Global Warming*

"By doing nothing, we risk the planet's future."

—SENATOR TIMOTHY WIRTH
*The Challenge of Global
Warming*

"Humanity is conducting an unintended, uncontrolled, globally persuasive experiment whose ultimate consequences could be second only to nuclear war."

—Conference on The Changing Atmosphere: Implications for Global Security, Toronto, 1988
The Challenge of Global Warming

"As there are no quick fixes or easy solutions, we must gear up for the long hard job of figuring out how the earth system operates. . . . Even with great intensification of effort, I fear the greenhouse impacts may come largely as surprises."

—WALLACE BROECKER
The Challenge of Global Warming

The End of Nature
BILL MCKIBBEN
1989, Random House

Bill McKibben, in *The End of Nature,* properly laments the irreversible destruction of life on the only planet that we know tolerates life. McKibben points out how the fossil fuel, chemical, and automobile barons are rendering the planet uninhabitable—and lobbying governments into doing nothing to diminish the release of greenhouse gases that bring the threat of disastrous global warming. They fail to realize that addicting humanity to gasoline and CFCs cannot produce lasting profits.

Civilization's actions, preponderantly those of the age of science, technology, and hubris, are driving species of plants and animals to extinction at an unforgivable rate, and there is no way we can call them back. McKibben's description of the unraveling of the web of life is so awesome that no one can sleep through it. *The End of Nature* is so convincing that it left me feeling optimistic. It ranks appropriately with *Silent Spring* and *Sand County Almanac* and should be shelved alongside the family Bible and reread frequently.

But I do not recommend that you buy a copy. Buy two. And with a yellow marker, highlight one and send it to the President. Ask him please to peruse it and to let you know how he intends to heed the alarm it sounds, remembering to thank him for saying that he is an environmentalist. Then call a book party meeting of your neighbors (when did you last do that?) and plan how to thank him again when he overcomes the environmental failures of his administration.

And perhaps you'd like to write Bill McKibben, as I did, suggesting that he start work on a sequel, "The Return of Nature," listing what you and your immediate friends are going to do to give him something to write about.

—DAVID R. BROWER

"The computer models, however, project an increase in global average temperature as high as a degree Farenheit per decade. An increase of 1 degree in average temperature moves the climatic zones thirty-five to fifty miles north—that's why, when you drive from Atlanta to New York the vegetation that lines the highways changes. So, if the temperature was increasing a degree per decade, the forest surrounding my home here would be due at the Canadian border sometime around 2020, which is just about the time that we'd be expecting the trees from a hundred miles south to start arriving. They won't—half a mile a year, remember, is as fast as forests move. The trees outside my window will still be there; it's just that they'll be dead or dying."

"The invention of nuclear weapons may actually have marked the beginning of the end of nature: we possessed, finally, the capacity to overmaster nature, to leave an indelible imprint everywhere all at once."

"The point is, it's easier, and thus 'more fun,' to use oil for almost everything. Raking leaves, for instance. By 1987 Americans alone had paid more than a hundred million dollars to buy electric leaf blowers—machines that blow leaves around a yard, thereby replacing the rake."

The Toxic Cloud: The Poisoning of America's Air
Michael H. Brown
1988, Harper & Row

In *The Toxic Cloud* Michael Brown provides a fascinating account of what he terms the poisoning of America's air. It is a sad and frightening story, told with riveting clarity.

Most Americans would be surprised to learn that the release of dangerous compounds into the air by chemical plants and other industrial sources is entirely unregulated. Few are aware that there is absolutely no requirement that those handling dangerous chemicals take precautions to prevent their accidental release into our air.

This isn't at all what Congress had in mind when the Clean Air Act was passed nearly twenty years ago. In fact, in 1970, on the heels of the first Earth Day, Congress directed EPA to tightly and promptly regulate "hazardous" air pollutants.

The agency has avoided this mandate simply by refusing to recognize clearly dangerous chemicals as "hazardous." Only seven of the hundreds of dangerous air contaminants have been regulated by EPA over the past two decades. Among those chemicals that EPA has declined to consider hazardous are numerous substances formally listed as cancer causers by the National Toxicology Program of the U.S. Public Health Service, such as formaldehyde, chloroform, and carbon tetrachloride. Even more flagrant examples include phosgene, a nerve gas which killed thousands in World War I, and methyl isocyanate (MIC), the chemical that killed more than three thousand

and injured tens of thousands more when it was accidentally released in Bhopal, India, five years ago.

It's not that these chemicals aren't used or released into America's air supply. EPA studies show routine air toxic releases of staggering magnitudes. According to industry reports, more than 2.7 billion pounds of toxic air pollutants were released into the nation's skies in the year 1987 alone.

And this is only part of the problem. Accidental releases of air toxics also occur with disturbing frequency. EPA reports more than eleven thousand accidental air toxic releases between 1980 and 1987, an average of more than two a day. The agency has documented seventeen accidents with a life-threatening potential comparable to the accident in Bhopal. According to EPA, only good fortune—the wind blowing in the right direction, the accident occurring at facilities not close to population centers—saved us from a catastrophic loss of life like that in Bhopal.

In *The Toxic Cloud,* Michael Brown depicts the sad story of personal and ecological tragedies from the nationwide air toxic assault. He describes how cancers, nervous system disorders, birth defects, and respiratory ailments have afflicted families from Staten Island to Louisiana to Michigan to California. We learn how a chemical industry out of control has turned neighborhoods into hot spots, where residents face inexplicably high disease rates and housewives and workers have learned, of necessity, to be environmental activists. And we learn how air toxics have been carried on the winds to the nation's rural areas, including even remote wilderness areas, and how they've built up in our ecological systems. Even prize game fish from remote areas of the Great Lakes are now considered inedible because of toxic contamination that could only have originated in the air.

The Toxic Cloud is a clarion call to protect the public and the planet from the toxic assault in our air supply. It is a message that we can ill afford to ignore.

—HENRY A. WAXMAN

"There is actually a sign, at a liquid petroleum off-loading gate, that says, 'DANGER . . . DO NOT DRIVE INTO VAPOR CLOUD.'

" 'Twelve thousand tons of these hydrocarbons, or unburned fuel, escape into the atmosphere each year in New Jersey simply from cars fueling up at gas stations,' an official complained to the Newark *Star-Ledger.* 'That translates to 4.5 million gallons of gasoline.'

"Fumes are also forced out of tankers loading up.

"A state official told me there once had been a test that showed the area near Newark International Airport to be the single most polluted spot in the nation. But it depends on what chemical you are looking for, and when the samples are taken. Benzene, in a run of samples by the New Jersey Institute of Technology, looked very high in the city of Elizabeth, but not as high as it was in Houston, Toronto, Phoenix, Tuscaloosa, or Los Angeles.

"But only in New Jersey was there an incinerator which malfunctioned and spat discarded birth-control pills onto a nearby street."

"EPA, at the bottom line, has been too lenient with industry.

"Indeed, sometimes it is difficult to tell if one is speaking to a federal official or a corporate salesman. Officials at the Louisiana attorney general's office told me they had taped a conversation with one EPA official who was trying to stop their opposition to a proposal by Rollins to burn PCBs in its problem-plagued incinerator. The official seemed to hint that if the state let Rollins have its way, Louisiana might be chosen as the site for a new naval base.

"Always more accommodating to corporations than individuals, no matter *who* is president, the EPA was in no hurry to expand its regulation of hazardous substances. In the seventeen years since enactment of the Clean Air Act—which, under Section 112, calls for control of toxic substances that might threaten human health—the agency had categorized only seven substances (arsenic, beryllium, mercury, benzene, vinyl chloride, asbestos, and radionuclides) as officially 'hazardous' and thus subject to regulation.

"At one point in 1984, by which time only 5 of 650 targeted chemicals were being regulated under Section 112, Congressman Gerry E. Sikorski of Minnesota commented, 'At that rate it will take 1,820 years to do something about the remaining chemicals on the list, almost as long as since when Christ walked on Earth.' "

"From preliminary indications, in fact, farm dust would appear to be a major, unrecognized source of toxic air pollution, dispersing weed killers and fungicides that can cause everything from relatively innocuous hay fever-like symptoms to fetal death. In one of the very few studies done along this line, researchers from the U.S. Department of Agriculture (DOA) found up to twenty thousand parts per trillion of herbicides and organophosphate insecticides in the fog hovering over two test areas: Beltsville, Maryland, and the Central Valley of California.

"Since the same readings could be expected anywhere in the expansive agricultural zones of the central states (and anywhere else, for that matter), the DOA findings should be heard as a clangorous warning.

"Dr. Dwight Glotfelty of the DOA informed me that the volume of farm chemicals evaporating directly into the atmosphere—not counting what escapes on windblown topsoil—can range from 'insignificant' levels to more than *half* of what is applied."

Fear at Work: Job Blackmail, Labor and the Environment
RICHARD KAZIS AND RICHARD L. GROSSMAN
1982; 1990, New Society Publishers

Although this book is nearly ten years old, it's as relevant as this morning's newspaper. Even as I was reading it toward the end of 1989, there were reports in the daily press about the economic need for offshore drilling off the coast of California; a proposal by the Occupational Safety and Health Administration that USX (formerly U.S. Steel) be fined $7.3 million for safety and health violations at two of its plants (seventeen workers have been killed at the plants since 1972); and results of a study showing that among thirty-three men who worked with an asbestos fiber that was used as a filter in Kent cigarettes during the early 1950s, twenty-eight have died (eighteen from malignant mesothelioma, lung cancer, and asbestosis) and four of the five surviving workers have asbestos-related diseases.

The human toll that environmental degradation takes is usually not tallied. What is tallied are industry estimates of the cost of environmental controls and regulations. And with those estimates comes the blackmail: "If we have to adopt these environmental standards, companies will disappear, jobs will be lost." Kazis and Grossman confront this argument head-on, marshaling facts and examples to show that business leaders usually don't know what they're talking about when it comes to estimating costs and that they conveniently neglect to mention the human and economic *benefits* that accompany a clean environment.

Fear at Work is a manual on how fighters for a better environment and fighters for working people can make common cause. The central issue, the authors point out, is, who benefits? They argue that environmental control really means a "redistribution of rights," shifting costs "from workers and the public to firms responsible for causing the damage." It's a neat book that even comes with this dividend: two chapters that give pithy histories of the two movements—the fight for environmental protection and the fight for workplace rights. Kazis and Grossman endorse this statement by a United Steelworkers official: "What we need is to get carcinogens out of our lungs and put good jobs in our community."

—MILTON MOSKOWITZ

"Employers and their supporters have been quite effective in restricting national economic and political debate to a fairly narrow set of 'alternatives' which do not challenge the primacy of business priorities, prerogatives, and power. They have convinced large segments of the public that there are *no* alternatives. People's beliefs can be shaken when they are told repeatedly that change is impossible, that their ideas are impractical. And they can be persuaded to 'be realistic' when employers make it clear that to persist might cost them their jobs."

"By maintaining the fiction that all economic growth leads automatically to more jobs and a richer society, corporate leaders justify pursuit of their private goals in the name of the public good. They use 'growth' as yet another variation of job blackmail. Steel companies assert environmental regulations are slowing growth and causing job loss, even though jobs in steel production were disappearing due to automation before the 1970 Clean Air Act. Oil and utility companies claim energy growth is needed to improve minority employment. Yet their own record in hiring minorities is weak: in 1978 minorities accounted for only 10 percent of total employment and less than 3 percent of managerial staff in oil and gas extraction and in the electric power industries."

"Environmental protection creates jobs. It also *saves* jobs. Fishing, forestry, tourism, agriculture, and the growing leisure and outdoor recreation industries are all important sources of jobs which depend directly upon clean water, clean air, and wilderness for their continuation and growth. Although neither industry nor government has tried to estimate the number of jobs saved as a result of the preservation of environmental quality, it is likely that many jobs would have been eliminated in these industries had the environmental legislation of the past decade not been enacted and enforced."

Crossroads: Environmental Priorities for the Future
EDITED BY PETER BORRELLI
1988, Island Press

The truly amazing thing to me about this book of essays, many of which were written by dedicated and well-meaning environmental folk, is that no one addressed head-on and systematically the question, how do we prevail?

The environmental movement is a political movement, isn't it? And the ultimate purpose of political movements, as I understand it, is to make governments do what you want them to do. After you decide what you stand for, there are of course all sorts of ways of going about this—grassroots organizing, nonviolent resistance and civil disobedience, coalition building, mass media appeals, electioneering.

Mainly what the established environmental organizations do today is lobby. But as Earth First!'s Dave Foreman points out, "we're in the middle of the greatest biological catastrophe since the dinosaurs died off 60 million years ago." Given that reality, rearranging the deck chairs on the Titanic, that is, lobbying, hardly suffices.

The citizens of Poland, Hungary, East Germany, and Czechoslovakia have given new life to the phrase "consent of the governed." It very well might take a comparable awakening of the American citizenry before their government takes decisive actions to preserve life on our planet as we know it. Given that this is the same citizenry that gave Ronald Reagan the highest approval rating of any president since FDR, the same citizenry that spends, on average, three hours a day watching television, it may take some doing.

—DAVID SHERIDAN

"In the summer of 1987, about 1,500 Greens gathered in Amherst, Massachusetts, for their first national conference. The stage was set for the Greening of America, but by lunchtime of the first day, the meeting had dissolved into an unhappening, as shouting matches broke out among deep ecologists, feminists, animal liberationists, anarchists, antimilitarists, monkeywrenchers, and graying SDSers. There was little talk of national organization or coalition building."

—PETER BORRELLI

"There is no way to reorganize society along ecologically sound lines without directly challenging the powerful, politically conservative forces—more plainly speaking, the corporations—that now control the system of production."

—BARRY COMMONER

"At the heart of many of the most important conservation disputes is a desire to change the way society operates, and the long-term conservationist cause may be better served at times by risking the loss of everything in litigation than by compromising."

—FREDRIC P. SUTHERLAND
AND VAWTER PARKER

"Since those heady days in the late 1960s when the nation took two giant environmental steps forward, we have taken a step back. But we are still one giant step ahead, and the footprint is there in our laws. In a relatively short time, we will take another step and regain the lost ground."

—JOHN N. COLE

"Paper victories are tough enough to come by, but they create illusions rather than effective environmental progress."

—MICHAEL FROME

FIRE

PROVIDING WONDROUS RELIEF from the appalling news of global warming and oil spills during the summer of '89 was the Voyager fly-by of Neptune. Once again we were reminded of the vast distances that separate Earth from the other planets and moons of our little solar system. Neptune, 2.79 billion miles away, was disclosed as an uncanny place indeed. For *our* planet to be just the right distance from the sun, to be reached by its light in just the right measure, to be warmed into life and neither baked nor frozen into an eerie abiological silence or chaos, is a remarkable happenstance. A perfect sunny location is the happy fate of the Earth, our world, so unlike any other that we have seen. Yet there is fire at the heart of the Earth, just as fire is the heart of our solar system, and the heart and hearth of human culture.

Fire animates and transforms, frightens and fascinates. The human relationship with fire is unique to our species. Humans found fire long before they made fire—perhaps a lightning-struck wildfire, a nervous red line trailing billows of smoke, chewing its way across the savanna. It's definitive that humans are fire-using animals, tool-using animals. Fire can cook meat, bake bread, glaze a pot, forge steel, and keep the beasts at bay. The campfire is an intimate sun in the night, dispelling the dark, consuming the fuelwood, leaving little in the morning but black coals and gray ash. Even in people to whom the sight of a fire in a circle of stones or in a hearth is rare, there is some deep knowing that this mastery is the essence of our power as a species. Fire has been one of our most ancient tools, a means of keeping openings for game, of clearing little plots for slash-

and-burn agriculture, and of creating prairies where favored wild plants could dependably be harvested. By and by came fire broiling a planked salmon under wet lowering skies. Next an ember from an old home coming to sanctify a new hearth. Then woodsmen going for the fire locked in ancient oaks. And now fire unleashed from anthracite coal, from crude oil, from fuels laid down millions of years ago, those famous fern forests abuzz with their three-foot dragonflies. Fossil fuels yielding with fire so hot it can melt steel.

Momentous changes came with our capturing of fire. Invention became a prime activity steadily gathering momentum, but it wasn't until the development of the steam engine, around 1769, that fire could be translated into motive power. Prior to that time, wind and falling water had powered mills, but those energy sources, although renewable, weren't portable, or necessarily constant. Thus the steam engine was a technological breakthrough of inordinate magnitude. About sixty years later the invention of the locomotive followed and inaugurated our modern era of transportation, centralization, and haste. Our burning of nonrenewable fuels has accelerated wildly over the last two hundred years, especially during the last fifty years, not merely changing the face of the Earth (and, with it, human culture) but also the very composition of the atmosphere.

We've come a long distance from those tentative, chancy beginnings of our relationship with this element. Fire still threatens to consume us—or rather by our use of it we threaten to consume everything. Today the fire in the chambers of our internal combustion engines is perhaps the single greatest culprit, although the technologies that preceded the automobile literally paved the way. Perhaps the worst thing we've done with fire has been the greenhouse effect. Primarily from our burning of fossil fuels, we are adding carbon dioxide (and several other "greenhouse" gases) to the atmosphere. (For instance, an automobile driven ten thousand miles a year will produce its own weight in CO_2 annually.) This increased concentration of greenhouse gases is trapping more solar warmth. The consequences of this global warming will beggar the imagination, and only radical change in our patterns of energy use can mitigate them.

Step one in reforming our use of energy is awareness taken right back to the campfire, remembering that what is burned is consumed, and that the smoke, given the right conditions, can hang like a pall over the valley; remembering that excessive brushcutting can kill the forest. This is very old knowledge, all but forgotten now. We can reclaim it and begin to make decisions based on the awareness that what you burn you breathe, and that wasting fuel or heat is worse than foolish.

The fact that the most spectacular environmental disasters in recent memory—global warming, acid rain, forest death, Chernobyl, and the Alaskan oil spill—have resulted from our seemingly insatiable demand for electric or mechanical power is good reason to eliminate the frivolous from our patterns of energy use. We can begin by asking ourselves, is this device, or this trip, or this product (for all our purchases are tokens of energy consumed) really necessary? Is it good for the Earth? Is it making me any stronger, healthier, or wiser? Could I do without it?

In the realm of fire, as with Earth and air, the individual, of course, has some effective choices to make when it comes to energy use. Learning how to be clever and frugal with energy, and mindful of the consequences of all combustion for the planet's atmosphere, can no longer be the province of experts. It is something every householder and apartment dweller and car owner can become involved in, and little by little, gain command over this pervasive factor of contemporary life.

The basic outlines and techniques for a sustainable energy economy have been understood for a decade now. Not much has changed except that the costs of some energy alternatives have come down, and the consequences of not changing our patterns have become very much more serious. Much of what needed to happen in order for us to make a graceful transition to more sustainable energy use was known two decades ago, but on account of a decline in oil prices and a failure in leadership, most people lost the incentive to conserve. Now we must arrive at bases for personal decision making other than self-interest defined in economic terms. Limiting ourselves to that is a degrading confession of choicelessness. And choicelessness, at this point, is a stance that we must adamantly reject.

Just how badly do we want to survive? And how strongly do we feel some obligation to leave a world for future generations? Can we want to passionately enough to begin changing the patterns of our lives? Of course. So much of the willingness starts with simple information.

Behavior counts. Usually behavior counts way more than belief. You can believe that the environment is in crisis and that it's important and that something should be done, but if none of the doing is taking place in your personal sphere, then nothing is changing. Because the planet is, in certain senses, finite, and somewhat of a closed system, we all live intimately with the results of our physical acts. Things do add up, and as population grows, there are more of us adding to the adding of things. Similarly, the benefits from many individual actions of self-restraint, frugality, and material simplicity will add up. It won't be easy, but it will be necessary.

Plant geneticist/agrarian philosopher Wes Jackson has observed that "we live in a time of structural immorality." The decisive role of the automobile in our society's assault on the biosphere is a perfect example of what he means. To avoid automobile use virtually requires stepping outside the social structure. If you don't happen to live in a city with reasonable mass transit, you have to be something of a vagabond saint to live without a car. What's a suburban commuter to do? A country dweller? A single woman wary of buses at night? Clamoring for more fuel-efficient cars is an interim step. And maybe market demand could lead to the production of affordable cars that will run on a fuel that won't produce carbon dioxide as a by-product. Hydrogen-powered or photovoltaic cars also might ease the transition. But then we'd still confront the structural problem of urban design and land use as determined by the car. The postwar pattern of energy use has led to a radical dependency on long lines of supply. Our food and fiber are grown far from the cities; our fruits and vegetables tend to be very well traveled. Other necessities— tools and nonrenewable fuels—also may be imported from other regions or countries. The result of this dependency has been an increasingly unsustainable level of urbanization and an abandonment of the countryside. Two-thirds of the open space in Los Angeles, for instance, goes to accommodate the automobile.

There are still places on Earth where it works other ways. In China the ratio of bikes to cars is 540 to 1. There's one bike for every four people. There, bikes are the main means of personal transportation, sustainable indefinitely, clean, health-promoting, moving at a speed that doesn't accelerate life unreasonably. Bikes represent the most energy-efficient traveling machines ever devised by humans. Bikes also impose a humane boundary on commutes and communities, not shattering social relations and the physical environment as radically as automobiles. Bikes definitely deserve the right-of-way in all our considerations of what to do about global warming, just as cars and all other energy-hungry technologies must forfeit the benefit of the doubt. Clearly bikes are not for everyone—the very elderly and infirm, for example—but given congenial urban design, they could be for more of us than we might presently imagine.

Between 1949 and 1973 energy consumption grew three times as quickly as population. If it grew that fast, why couldn't it decline that fast? Fossil-fuel burning for transportation and electrical generation demands extractive activity that results in horrors like strip mining and offshore oil spills. Combustion of fossil fuels and, to a lesser extent, the burning of tropical forests are leading to global warming. Coal burning leads to acid deposition, now severely affecting forests throughout the Northern Hemisphere.

The hose that fills my gas tank once or twice a week is my direct tie to the events in Prince William Sound. Every time I gas 'er up, I am taking on some participation in seabird and otter death, some responsibility for the reduction in the number of eagles and herrings in a great arctic ecosystem. I pardon my goofy little trips to town, or to the library, just me alone in my sedan, getting not very good mileage, and doing things that could be accomplished in part by better planning (stocking up in advance) and greater self-reliance (a more productive garden and putting food by). I could go to a desktop data base instead of a library. And I could go to every county commission meeting and plead for improvements to our regional transit system so that there would be a better alternative to the one-owner passenger car mode in this far-flung neosuburb of ours. Not to acknowledge that there's much more that I can be doing, but blaming the system, and continuing not to choose to reduce or eliminate my personal gasoline consumption is like blaming my bad behavior on my parents, and excusing it, long after I cruise into middle age.

Yet it is amazing, given the global consequences, that the manufacturers of fuel-inefficient cars don't retool immediately and pledge themselves to another course. They have choices too. As stately a publication as the *Economist* quipped, "There is more oil under Detroit in the form of untapped fuel efficiency than in the Alaska field, America's biggest reserve." Efficiency improvements alone could save twice the officially anticipated oil supply from the Alaskan fields, and there is no telling what savings could be accomplished through a program of abstinence from many wasteful forms of consumption of petroleum.

Extraction of energy resources has always exacted a heavy toll from the Earth, beginning with the deforestation of the Mediterranean basin, not just to build ships, but to melt the metals that named the Bronze Age. Coal mining was another killer, first drawing men deep into the Earth, into dangerous toil, and then in our time, with the advent of strip mining, scraping bare whole counties and sacred sites to get at fossil sunlight. Uranium extraction for the nuclear economy displaces traditional peoples, exposes miners to deadly radiation, and inaugurates a wildly expensive energy-generating system whose by-products are toxic for millennia.

Given the costs and hazards of large-scale energy generation, the big buzzword in energy policy these days is "least-cost planning." It means weighing the true costs—including environmental impacts—of various means of generating energy against one another and counting conserved energy as a source. It's the energy version of "a penny saved is a penny earned." Least-cost planning has in some cases showed that it's considerably cheaper to conservation retrofit a whole region's houses than to build a new electrical power plant. What's more, enduring benefits are realized by the one-time expenditure for conservation measures, while the detriments and expense of producing power by combustion and fission escalate continually. If global warming weren't enough to persuade us to phase out the burning of fossil fuels, there remains the fact that on a finite Earth, nothing exists in unlimited supply. Sooner or later we will run out. With the case of Alaskan oil, in light of the dreadful environmental consequences of removing and transporting it from that fragile environment, why couldn't we just act as though it were gone already?

The sun beams ample energy to planet Earth, in a form that is supremely useful to a lot of living organisms—the plants. Solar light makes grass and flesh. It "causes" photosynthesis, that primal source of all fuels but nuclear. Sunlight can heat a house in winter and charge a battery to run a fan in the summer. Environmentalists have long been fond of saying that the sun is the only safe nuclear reactor, situated as it is some ninety-three million miles away. The Earth's orbiting the sun, the Earth's tilt of axis, vary the length of our days, change the seasons, drive the climate—wind and ocean currents. Earth's dynamic all proceeds from the sun. Our ancestors understood that and were far more sensitive than we to what they perceived as the sun's comings and goings, observing a yearly cycle of holidays at the solstices, equinoxes, and cross-quarter days. Nowadays alertness to the sun's track through the sky can be indispensable to someone siting a home to take greatest advantage of solar light and warming, or building a greenhouse, or locating a garden.

The hopes for alternatives, for renewable energy, are as gentle and persistent, as constant and illuminating of our better natures, as the sunlight, a gentle breeze, or falling water, all of which can, utilized at the correct scale, supply renewable energy. During the first golden age of solar research and development, renewable sources for large-scale power generation began to be explored, with wind farms being the most promising. Other modes of centralized power production—Ocean Thermal Energy Conversion (OTEC), solar furnaces, and mass photovoltaic collection—seem less so. Fortunately, it is possible to design and engineer the dwellings, tools, and vehicles necessary to help people meet their basic needs utilizing sustainable flows of energy from a diversity of sources, each suitable to a particular end use. The whole appropriate technology movement, whose mission this is, affirms and moves with the grain of human ingenuity, offering myriad ways to ease the toil of meeting basic needs. More sophisticated high technologies for monitoring, reducing, and refining energy consumption hold great promise. Architecture and urban design can once again follow the track of the sun and defer to the weather to reduce the need for energy to provide heat and light, and to restore the strength and joyousness of human mobility, by foot or pedal power. Also, moving toward simpler, less wasteful household economies with all due speed will help greatly.

Changing energy consumption in the home gets us right where we live. Life in temperate regions without home heating (and, in an era of global warming, air conditioning) would be essentially impossible. However, energy maven Amory Lovins talks of the waste and futility involved in "trying to heat a sieve," implying that there are numerous ways, requiring different degrees of planning and expense, to tighten up the sieve and reduce energy loss from a home. The possibilities range from caulking to stop drafts, and tree planting for shade and windbreaks, to elaborate solar architecture, super-efficient appliances, envelope houses, photovoltaic and wind systems, and the like. Supplying energy through conservation is in myriad details, such as keeping the lids on your pots, keeping the pot centered over the flame, and making the shift to energy-conserving lightbulbs. According to the Rocky Mountain Institute, an 18-watt compact fluorescent lightbulb, which will give as much light as a 75-watt incandescent bulb and last thirteen times as long will, over its lifetime, "slow global warming and reduce acid rain by avoiding emissions from a typical U.S. coal plant of 1,000 pounds of carbon dioxide and about 20 pounds of sulphur dioxide. . . . The same fluorescent bulb will also save the cost of buying and installing a dozen ordinary bulbs . . . the cost of generating 570 kWh of electricity . . . and, during its lifetime, approximately \$200 to \$300 worth of generating capacity [incurred by the utility company]." So even making an almost-imperceptible change at the household level can produce tangible environmental and economic benefits. And, once begun, this change can generate its own momentum.

Some of us are lucky to live in locales where a significant number of householders have taken up the challenge of ecological transformation toward a saner tomorrow. For instance, I know a guy who has been responsible for getting more than two million trees planted. Where keeping homes warm in the winter is challenging and generally expensive, there's a strong incentive to minimize these costs. In my county there are quite a few people who are almost competitive about their energy conservation measures, their Earth-restoring activity, and their efforts toward self-sufficiency. Among my friends are three couples who have, with their own labor and not much money, built houses that are completely independent from the power grid. These people are continually swapping their intelligence and techniques. So far as I know, no institution taught them. They put their systems together with the help of dog-eared catalogs from little seat-of-the-pants manufacturers and entrepreneurs doing business out of high-mountain villages. My friends are not engineering geniuses or saints; they're just people who got the message and felt both responsible and empowered to do something about it. They found ways to plan and build for sustainability and harmlessness. And if they can at the homestead level, you can. I can. Even the timid can begin to take steps in the direction of greater household energy efficiency by small deeds, like converting to compact fluorescent lightbulbs where possible (and it is in most fixtures). For urban dwellers, that planning and building will need to take place in the social sphere first—neighborhood organizing, car pools, cooperatives of various kinds, and working with municipal governments to subsidize conservation retrofits for low-income families, for instance. We could also respectfully suggest that there need not be a blazing light for every broken heart on Broadway or in Las Vegas and agitate for a less gaudy, more elegant, and energy-conscious aesthetic in our cities.

In the fairly recent past "passive solar" energy application (structural design to capture solar light and heat by house siting, Trombe walls, rooftop solar collectors for hot water heating, or greenhouse construction) was all the rage. There are also such time-honored passive solar techniques as reading by daylight and drying clothes on a line on a sunny day. There seems to be nothing passive, however, about the current generation of photovoltaic homesteaders. They are learning how to wire their own homes to accept photovoltaic and wind energy as delivered (or gathered) and are becoming connoisseurs of an ever-growing array of twelve-volt appliances and energy-efficient lightbulbs and pumps. In order to take advantage of these elegant new technologies for self-reliance, they are starting from scratch and gaining mastery over their own energy systems.

For city dwellers, tenants, or owners unable to start from scratch in energy-system design, steps towards sustainability and harmlessness may seem less dramatic. Purchasing decisions would be important—seeking out the most energy-efficient version of the item, be it a car, refrigerator, or small appliance, for instance. Perhaps more important would be nonpurchasing decisions—choosing to do without energy-consuming gadgets, not giving them as gifts. Frugality might become the latest style, simplifying life by gaining mastery over one's wants. Such action, combined with lively civic participation, aimed at "greening" the city, is a worthy task for the creative intelligence, and very human—like gaining mastery over fire.

To the argument that it isn't practical for us to try to effect such sweeping cultural and technical change, we can simply respond by asking how practical will it be for us to bumble our way through climate change the like of which hasn't been seen since the last glaciation, but which will be transpiring over decades rather than millennia? How tough a choice is it, really?

The Road to Wigan Pier
GEORGE ORWELL
1937; 1972, Harcourt Brace Jovanovich

In the mid-thirties, the Left Book Club's board of editors asked George Orwell for a documentary report on conditions among unemployed workers in northern England. As the editorial foreword (retained in the American edition) makes clear, Orwell wildly exceeded his assignment—to the quibbling dismay of the editors, but to the delight of this reader.

The first half of *Wigan* is, as commissioned, a look at working-class conditions in the industrial North, but it examines the employed as well as those "on the dole," for as Orwell discovered, work and welfare were equally oppressive. His extended description of the daily lives of coal miners is detailed with such merciless clarity you can taste the coal dust, and feel it settle in your lungs. His portrait of the working class as virtual slaves is brutally convincing, and not only on the expected terms of class analysis—Orwell is among the first to note that the reduction of human beings into economic units occurs simultaneously with industrialism's pillage and poisoning of the environment, and while one might argue that the biocidal impulse existed before the Industrial Revolution, fossil fuels increased the range and rapidity of destruction as well as adding their own toxic by-products to the biospheric dump. Orwell's portraits of the fouled skies and ravaged landscapes are dismally vivid.

If Orwell had ended there, letting his descriptions indict the oppressors (the usual consortium: military/state/big biz/church and their insulting use of the Earth), the editorial board might have been spared untangling the party line. But the first half of the book, marked by the sickening accretion of "objective" details that convey their own wretchedness, indignity, and exhaustion, is followed by an intimate account of Orwell's middle-class background, most pointedly his five years' government service with the Indian Police, which eventually gathers into a passionate "class analysis," and concludes with an attack on socialism's failure to inspire much more than disdain from its theoretically natural allies, the workers.

The Road to Wigan Pier has some wild curves and a few missing bridges, but if you hang on and make the leaps, you'll find the full dimension of Orwell's powers: lucid and graceful exposition; analysis that is thorough, honest, relentlessly precise; and a moral imagination whose indignation is tempered by compassion.

—JIM DODGE

"As you travel northward your eye, accustomed to the South or East, does not notice much difference until you are beyond Birmingham. In Coventry you might as well be in Finsbury Park, and the Bull Ring in Birmingham is not unlike Norwich Market, and between all the towns of the Midlands there stretches a villa-civilisation indistinguishable from that of the South. It is only when you get a little further north, to the pottery towns and beyond that you begin to encounter the real ugliness of industrialism—an ugliness so frightful and so arresting that you are obliged, as it were, to come to terms with it. . . .

"When you contemplate such ugliness as this, there are two questions that strike you. First, it is inevitable? Secondly, does it matter?

"I do not believe that there is anything inherently and unavoidably ugly about industrialism. A factory or even a gasworks is not obliged of its own nature to be ugly, any more than a palace or a dog-kennel or a cathedral. It all depends on the architectural tradition of the period. The industrial towns of the North are ugly because they happen to have been built at a time when modern methods of steel-construction and smoke-abatement were unknown, and when everyone was too busy making money to think about anything else. They go on being ugly largely because the Northerners have got used to that kind of thing and do not notice it. Many of the people in Sheffield or Manchester, if they smelled the air among the Cornish cliffs, would probably declare that it had no taste in it. But since the war, industry has tended to shift southward and in doing so has grown almost comely. The typical post-war factory is not a gaunt barrack or an awful chaos of blackness and belching chimneys; it is a glittering white structure of concrete, glass and steel, surrounded by green lawns and beds of tulips. Look at the factories you pass as you travel out of London on the G.W.R.; they may not be aesthetic triumphs but certainly they are not ugly in the same way as the Sheffield gasworks. But in any case, though the ugliness of industrialism is the most obvious thing about it and the thing every newcomer exclaims against, I doubt whether it is centrally important. And perhaps it is not even desirable, industrialism being what it is, that it should learn to disguise itself as something else. As Mr. Aldous Huxley has truly remarked, a dark Satanic mill ought to look like a dark Satanic mill and not like the temple of mysterious and splendid gods."

Technics and Civilization
LEWIS MUMFORD
1934; 1963, Harcourt Brace Jovanovich

It would be foolhardy to name any single person as the first environmentalist, but let's be foolhardy.

My nomination is Lewis Mumford, and I'm prepared to defend my view that he was not only the first environmentalist but that his twenty-eight books represent the most profound and comprehensive thought on the relation of humans to their total environment, both cultural and natural.

The conservation movement, with John Muir as the pioneer, had long fought valiantly for the preservation of wilderness and parklands, but there was little public recognition of the environment as a totality until the late 1960s, thirty-five years after Mumford's first writings on the subject.

Technics and Civilization, which he finished in 1931, was a resounding trumpet call, as auspicious a beginning as the early essays of Emerson or the first edition of *Leaves of Grass.* It led to a magisterial series of books, well-nigh astounding in scope, literary brilliance, and prophetic insight. Histories of technology had been written before, but they were almost entirely the stories of mechanical contrivances. *Technics and Civilization* broke new ground in looking at the impact of technology on a society that had become increasingly devoted, during the Industrial Revolution, to the culture of machines rather than the nurture of life.

In this book Mumford unearthed a forgotten classic, George Perkins Marsh's *Man and Nature* (1866), which called attention for the first time in a systematic way to the destruction of the forests and the erosion of the soil as elements in the decline of civilizations. He also anticipated the concept of Gaia: "Instead of accepting the Victorian myth of a struggle for existence in a bland and meaningless universe, one must . . . replace this with the picture of a partnership in mutual aid, in which the physical structure of matter itself, and the very distribution of elements on the earth's crust . . . are life-furthering and life-sustaining."

Instead of patchwork political reforms Mumford called for a wholesale reorientation of values, from the mechanical to the organic; from quantity, measured in production units or dollars or weapons, to the quality of life, which is immeasurable and intuitive; from exploitation of nature to cooperation; from domination to community.

His works offer perhaps this century's most sweeping vision of our historic era, its sickness and health, its unprecedented dangers and its unsurpassed opportunities.

—HAROLD GILLIAM

"In this paleotechnic world [the Industrial Revolution] the realities were money, prices, capital, shares: the environment itself, like

104

most of human existence, was treated as an abstraction. Air and sunlight, because of their deplorable lack of value in exchange, had no reality at all. . . . So this period was marked throughout the Western World by the widespread perversion and destruction of environment. . . ."

"With a change of ideals from material conquest, wealth, and power to life, culture, and expression, the machine will fall back into its proper place: our servant, not our tyrant."

"Equilibrium in the environment . . . means first the restoration of the balance between man and nature. The conservation and restoration of soils, the re-growth, wherever this is expedient and possible, of the forest cover to provide shelter for wild life and to maintain man's primitive background as a source of recreation, whose importance increases in proportion to the refinement of his cultural heritage. The use of tree crops where possible as substitutes for annuals, and the reliance upon kinetic energy—sun, falling water, wind—instead of upon limited capital supplies. The conservation of minerals and metals; the large use of scrap metals. The conservation of the environment itself as a resource, and the fitting of human needs into the pattern formed by the region as a whole. . . ."

"In order to maintain the ecological balance of a region, one could no longer exploit and exterminate as recklessly as had been the wont of the pioneer colonist. The region, in short, had some of the characteristics of an individual organism: like the organism, it had various methods of meeting maladjustment and maintaining its balance: but to turn it into a specialized machine for producing a single kind of goods—wheat, trees, coal—and to forget its many-sided potentialities as a habitat for organic life was finally to unsettle and make precarious the single economic function that seemed so important. . . ."

THE FOREST AND THE TREES

During the seventies, when I worked as a tree planter in the Pacific Northwest, I kept a tattered print of the seventh-century Buddhist priest Hung-jen above my writing desk. In monk's robe and sandals, the old priest is pictured climbing a mountain trail with a sack of tree seedlings and a planting hoe over his shoulder.

For many of us living in the rural Northwest, planting trees seemed the right thing to do. It brought us out early in the year, as winter snow melted back from the foothills. The air was newly washed, the mountain streams were clear, and the rhythmic pace of the work allowed plenty of room for friendly talk. When we stopped to look back over a slope we had crossed, fresh young seedlings spread out behind us into the misty green distance.

Planting trees is something everyone can do. It enables us, in a small way, to give something back to the Earth. The deep bow that accompanies the planting of each tree is a gesture of hope. And the life of the tree becomes a part of our own lives. Trees lend to us a certain grace and resilience, and their growth and maturity mirror our own.

As the pace of logging accelerated through the mid-seventies, and clear-cuts pressed into higher, steeper, more marginal ground, it became obvious that replanting was not enough. If we truly cared about forests and the future of forestry in our region, then the work of planting had to be complemented with the more difficult work of citizen activism.

For Hung-jen and his fellow monks it was a simple matter: All life was sacred. Even so, they were planting their trees during a time of massive deforestation throughout much of China and Japan. The soil erosion, drought, and desertification that resulted had devastating effects on the Asian landscape that have persisted into this century. The same pattern of deforestation, often followed by overgrazing and poor agricultural practices, has left barren landscapes in the West as well, from Egypt and the Mediterranean basin to the Oklahoma dust bowl.

It was into such a landscape that the memorable character of Elzeard Bouffier came in 1913, and began planting trees. Jean

Giono's inspiring story *The Man Who Planted Trees but Grew Happiness* serves as both allegory and anthem for the movement to restore the Earth. Bouffier's historic counterpart may well be the late Richard St. Barbe Baker. His autobiography, *My Life, My Trees,* recounts the author's lifelong effort to establish an international movement dedicated to conserving and replanting the world's once-forested areas. Forests standing today in India, Africa, Asia, and his native Britain attest to St. Barbe Baker's vision. The work of his organization, Men of the Trees, is carried on in North America by Friends of the Trees. Their 1988 *International Green Front Report* is a useful compendium of organizations and information concerning the regreening of the Earth.

In his 1929 classic, *Tree Crops: A Permanent Agriculture,* J. Russell Smith saw food trees as the obvious answer for farming on hilly or marginal lands. Crop trees could provide much of the world's food needs while conserving our most vital and basic resource, soil. On a contemporary note, Gordon Robinson's name should be added to the ranks of these earlier visionaries. Robinson's *The Forest and the Trees: A Guide to Excellent Forestry* is a careful and informed call for a return to true multiple-use forestry in our national forests. It provides a reasoned critique of current forestry practices, as well as an informative handbook for citizen involvement in forest policy decision making.

In this time of vanishing forests throughout the world, the thousand-year-old image of Hung-jen often returns to me. It represents the power and vision of the human spirit in the face of overwhelming odds. Today, that may be just what sees us through.

—TIM MCNULTY

The Man Who Planted Trees but Grew Happiness
JEAN GIONO
1967, Friends of Nature*

My Life, My Trees
RICHARD ST. BARBE BAKER
1970, Findhorn Press

International Green Front Report
1988, Friends of the Trees†

Tree Crops: A Permanent Agriculture
J. RUSSELL SMITH
1929; 1987, Island Press

The Forest and the Trees: A Guide to Excellent Forestry
GORDON ROBINSON
1988, Island Press

"On the site of the ruins I had seen in 1913 now stand neat farms, cleanly plastered, testifying to a happy and comfortable life. The old streams, fed by the rains and snows that the forest conserves, are flowing again. Their waters have been channeled. On each farm, in groves of maples, fountain pools overflow on carpets of fresh mint. Little by little, the villages have been rebuilt. People from the plains, where land is costly, have settled here, bringing youth, motion, the spirit of adventure. Along the roads you meet hardy men and women, boys and girls who understand laughter and have recovered a taste for picnics."

—*The Man Who Planted Trees but Grew Happiness*

* Friends of Nature, P. O. Box 223, Brooksville, Maine 04617
† Friends of the Trees, P.O. Box 1466, Chelan, Washington, 98816

Table of Contents

—International Green Front Report

"Forest—field—plow—desert—that is the cycle of the hills under most plow agricultures. . . . Indeed we Americans, though new upon our land, are destroying soil by field wash faster than any people that ever lived—ancient or modern, savage, civilized, or barbarian."

"As the deep-rooting, water-holding trees show their superior crop-producing power in dry lands, we may expect some of our now-arid lands to become planted with crop trees. Thus by using the dry land, the steep land, and the rocky land, we may be permitted to increase and possibly double our gross agricultural production and that, too, without resort to the Oriental miseries of intensive hand and hoe labor."

—*Tree Crops*

"Of the 30 billion acres of land on earth, more than 9 billion are already desert. We cannot afford to lose more of this green mantle, or the water table will sink beyond recall. Trees are like the skin of the earth. If a being loses more than one-third of its skin, it dies. One third of every country should be kept in tree cover.

"The country's very poor that doesn't have trees!"

—RICHARD ST. BARBE BAKER,
source unknown

"In this book I am deliberately offering a polemic in support of a return to uneven-aged management of our national forests—management in which stands of trees within a forest are small and irregular in size and shape, trees are of a variety of ages and species and are allowed to grow to maturity before being logged, and logging is done selectively and with care. This type of management safeguards the rich variety of trees and vegetation within a forest. It also protects all the forest's uses simultaneously and in perpetuity: the soil, watershed, fish and wildlife, and aesthetic beauty, and thus its value for outdoor recreation as well as its ability to supply a sustained yield of high-quality timber."

—*The Forest and the Trees*

110

Rays of Hope
Denis Hayes
1977, W. W. Norton and Company

As a cause of environmental problems, energy has few peers. The search for, extraction of, and use of energy fuels is at the root of our most pressing environmental problems, including oil spills, air quality, much tropical deforestation, weapons proliferation, and climate change. Reducing overuse of energy is a terrific way to solve such problems.

But look around in a bookstore. The most likely place to find an energy book today is under "climate change," which itself may be hard to find, squeezed in on one side by chaos and fractals and on the other by ADVANCED MS-DOS.

Once found, most climate-change books explain we will soon be living in a sickly, sticky greenhouse unless we curb excess energy use. The trouble is that few, if any, recent books lay out a plan and context for solving the problem. A classic one to begin with is this thirteen-year-old gem from the founder of Earth Day.

Nicest about this book is its enduring common sense unencumbered by the latest in knee-high footnotes. Hayes clearly explains the dangers of excessive dependence on fossil fuels, the risks of shifting to nuclear power, and the promise of energy conservation and new energy sources. Not a bit of the argument is timeworn or overblown. The short but solid chapter on nuclear power is good medicine for those who want to give it a second chance.

On the other hand, readers may have to fill in some of the last decade themselves. Hayes misses the fall of oil and gas prices, the quiet but resounding success of energy conservation, the worldwide plunge in nuclear power expectations, changes in expectations for solar sources, the energy and environmental politics of the Reagan era, and the new urgency imposed by climate change.

Seek it out; if you agree that some of the numbers need refreshment, join me in urging an update on its author and publisher.

—Jim Harding

111

"The energy crisis demands rapid decisions, but policies must nevertheless be formulated with an eye to their long-term implications. In making each of hundreds of discrete decisions, we would be well advised to apply a few basic criteria. Thrift, renewability, decentralization, simplicity, and safety should be the touchstones. Using these, we might judge whether a given action will move us closer to, or further from, the type of energy system we ultimately seek.

"Both rich industrial countries and poor agrarian ones can cull far more benefits in the immediate future from investments in increased efficiency than from investments in new energy sources. In fact, because they are unable to afford to make the necessary initial investments that conservation sometimes requires, the poor frequently waste a higher fraction of the energy they use than do the well-to-do. By eliminating waste and by matching energy sources carefully with appropriate uses, people can wring far more work from every unit of energy than is now the case. A sensible energy strategy will help accomplish this sensible goal.

"Energy is a means, not an end. Its worth derives entirely from its capacity to perform work. No one wants a kilowatt-hour; the object is to light a room. No one wants a gallon of gasoline; the object is to travel from one place to another. If our objectives can be met using a half, or even a quarter, as much energy as we now use, no benefit is lost.

"Investments in conservation must mesh with plans for a rapid switch from fossil fuels to sustainable energy sources. An intelligent strategy will lead to dependence upon energy derived solely from perpetually reliable sources. Solar technologies alone can provide us with as much energy as can be safely employed on our fragile planet.

"In establishing priorities for the post-petroleum period, foremost attention should be given to basic human needs—to food, shelter, clothing, health care, and education. Fortunately, such needs either require comparatively little energy or have energy requirements that can be met with renewable energy sources. Indeed, for most of history *Homo sapiens* has been entirely dependent upon renewable energy sources, and could not have survived if renewable sources had not met the most basic needs."

112

Soft Energy Paths: Toward a Durable Peace
AMORY B. LOVINS
1977; 1979, Harper Colophon Books, out of print

Returning to *Soft Energy Paths* nine years after it first inspired me, I am struck by its durability. The critical framework Amory Lovins presented for thinking about energy issues still holds—what tasks require energy, in what forms, in what quantities, by whom, and how that energy can be obtained. In revisiting this classic, I felt that I was exhuming Lovins's roots—back then, he used economics as only one of several reasons to follow the soft (meaning diverse, decentralized, renewable) path; now his energy work hews to a free-market line. In *Soft Energy Paths,* he explicitly wrote about his beliefs and ideologies, instead of minimizing his differences with the mainstream. He credited others generously, not just for the numerical analyses he drew on, but for the conceptual foundation that he developed.

Some things have changed. The reliance in *Soft Energy Paths* on clean coal combustion seems anachronistic, especially with heightened concern over coal burning because of global climate change. We now know how to save energy much more cheaply than was then thought possible. Later Lovins books have more current facts—such as *Brittle Power,* in which he and coauthor Hunter Lovins argue to the national security establishment that renewable energy will make the United States less vulnerable to disruption.

But much remains the same. People don't need energy for its own sake; they need it to provide certain end uses. This seemingly self-evident notion transformed the energy debate and slew dozens of power plants. The way of examining energy issues that Lovins lays out in *Soft Energy Paths* has outlasted an oil shock, an oil glut, two U.S. administrations, and the Ayatollah Khomeini. It will persist no matter what the price of a barrel of oil.

—SETH ZUCKERMAN

"Civilization in the United States, according to some, would be inconceivable if people used only, say, half as much electricity as

113

now. But that is what they did use in 1963, when they were at least half as civilized as now. What would life be like at the per capita levels of primary energy that Americans had in 1910 (about the present British level) but with doubled efficiency of energy use and with the important but not very energy-intensive amenities that people lacked in 1910, such as telecommunications and modern medicine? Could it not be at least as agreeable as life today? Since the energy needed today to produce a unit of GNP varies more than one hundred-fold depending on what good or service is being produced, and since GNP in turn hardly measures social welfare, why must energy and welfare march forever in lockstep? Such questions today can be neither answered nor ignored.

"Underlying energy choices are real but tacit choices of personal values. . . . Those that make a high energy society work are all too apparent. Those that could sustain lifestyles of elegant frugality are not new; they are in the attic and could be dusted off and recycled. Such values as thrift, simplicity, diversity, neighborliness, humility, and craftsmanship—perhaps most closely preserved in politically conservative communities—are already, as we see from the ballot box and the census, embodied in a substantial social movement, camouflaged by its very pervasiveness. Offered the choice freely and equitably, many people would choose, as Herman Daly puts it, 'growth in things that really count rather than in things that are merely countable': choose not to transform, in Duane Elgin's phrase, 'a rational concern for material well-being into an obsessive concern for unconscionable levels of material consumption.' "

"There exists today a body of energy technologies that have certain specific features in common and that offer great technical, economic, and political attractions, yet for which there is no generic term. For lack of a more satisfactory term, I shall call them 'soft' technologies: a textural description, intended to mean not vague, mushy, speculative, or ephemeral, but rather flexible, resilient, sustainable, and benign. . . . The distinction between hard and soft energy paths rests not on how much energy is used, but on the technical and sociopolitical *structure* of the energy system, thus focusing our attention on consequent and crucial political differences."

114

RMI Newsletter
Rocky Mountain Institute (RMI)*

Eco-villains have always made tempting targets; it's so easy to casti-
gate them, so safe to offer glib solutions to the problems caused by
their unrighteous enterprises. Unfortunately, name-calling and
prickly confrontation most often serve only to make the sinner
smiters feel good, while the sinners persist in their wickedness.

Amory and Hunter Lovins were more interested in being *effective*
than in name-calling when they founded the Rocky Mountain Institute in 1982. Styled as "an independent, nonpartisan, nonprofit
resource policy center," RMI has become the world's most powerful
antidote to dangerous and stupid energy and resource mismanage-
ment. Figuring that the foe is ignorance rather than malice, RMI
researchers assiduously gather and analyze the numbers, often from
governments, the military, and large corporations—sources some
environmentalists would consider suspect. RMI findings usually indicate policy changes that will benefit the "enemy" as much as the
environment and the rest of society. What a better way to encourage
important changes in high places!

RMI offers consultation with governments at all levels, educators,
and just plain individual citizens. They publish a host of papers and
books, plus massive studies intended for governmental use. They've
done our homework for us. Their work works! To find out how you
can help, join and use RMI, write to them or give them a call at (303)
927-3851.

—J. BALDWIN

"When RMI launched our COMPETITEK energy information
service in early 1988, we did not know how it would fare in the
marketplace. To date, however, COMPETITEK has been a resounding success. In less than two years, more than 100 organizations,

* Rocky Mountain Institute, 1739 Snowmass Creek Road, Old Snowmass, Colorado 81654-9199

ranging from Southern California Edison to the USSR Academy of Sciences, have contracted to receive COMPETITEK's reports on advanced techniques for electric efficiency."

"Energy costs have a hidden yet deleterious impact on local economies. In most towns, 89 to 90 cents of every dollar citizens spend on energy leaves their community and never returns. The *Energy Casebook* describes how energy-efficiency projects have enabled many towns to elude this stark reality and 'plug the leaks.' For example, an energy management program undertaken by Fremont, Nebraska is saving citizens $95,700 annually—money that would otherwise be going to buy energy from Wyoming, Canada, and Saudi Arabia."

"Economic Renewal: Two Case Studies
"Wisconsin Power & Light Company (WP&L) provides shared-savings energy efficiency projects to local business and agriculture. The projects have a maximum five-year contract and are often financed as joint ventures of the utility, local banks, and the business being served. Two examples of the types of projects undertaken are a wood-waste cogeneration system and an energy retrofit of a dairy operation. This shared-savings program has been very successful and has attracted attention outside of WP&L's service area. Recently WP&L was approached by another utility that wanted to franchise WP&L's program. It seems the other utility had been told by some of its manufacturing customers that they were thinking about moving to Wisconsin so they could take part in WP&L's efficiency program."

—*RMI Newsletter,* winter 1989

A Green City Program: For San Francisco Bay Area Cities and Towns
WRITTEN AND EDITED BY PETER BERG, BERYL MAGILAVY,
AND SETH ZUCKERMAN
1989, Planet Drum Books*

Ecocity Berkeley: Building Cities for a Healthy Future
RICHARD REGISTER
1987, North Atlantic Books

The principles of ecology have to be brought down to Earth, where we live—which for better or worse is in cities, including cities that until recently have been considered suburbs. (And a lot of it *is* for the better, since city dwellers are far more energy efficient than rural or suburban people.) *A Green City Program: For San Francisco Bay Area Cities and Towns* is not just a few individuals' dream of what cities might be like in a Green society. It derives from a lengthy and elaborate series of meetings organized by the founders of bioregionalism, The Planet Drum Foundation, which involved both thinkers and doers. The result is a well-thought-through and detailed program, each chapter paying off with a section called "What can cities do to promote . . ." for topics including urban planting, lower-impact transportation, renewable energy, neighborhood empowerment, recycle and reuse, celebrating life-place vitality, promoting wild habitat, and socially responsible small businesses. Each chapter also has a "fable" dramatizing how citizen action can bring healthy change on a human scale. These, together with visionary "what's possible?" sections, bring the greening of cities within reach of ordinary people pursuing sensible goals upon which consensus should be possible. While the immediate aim of the book is to galvanize action in the Bay Area, its suggestions are valid and inspirational for any city.

Ecocity Berkeley is a trail-breaking book, essential reading for

* Planet Drum Books, P.O. Box 31251, San Francisco, California 94131

anyone concerned with the future of our cities. Register has taken an American city, which is in its physical fundamentals just like our other small cities, and shown convincingly how it could, over a span of perhaps fifty years, become a more compact, congenial, elegant, and civilized place—with a fraction of the impact on the biosphere, and fitting snugly into its niche in its bioregion. Register is an artist as well as an ecological thinker (and organizer—of the group Urban Ecology) and thus provides not only cogent maps of the evolution of an ecocity but also charming and inspiring perspective drawings of the result. He shows how ecological principles can generate civil policies, such as "access by proximity, not transportation," or three-dimensional density interspersed with greenbelts. His contention is that this process can be challenging enough to displace our national obsession with military competition and to absorb all the creative energy we can bring to bear on it. He also sketches tax and planning policies that could guide the transition toward ecocities. The vision expressed here is readily generalizable to any modern city, and the sooner we get on with it the better.

—ERNEST CALLENBACH

"Longer-term visions for municipal action

- Reduce the width of all but arterial streets and their maximum speed limits to make the streets more hospitable to pedestrians and cyclists.
- Design some streets to include cul-de-sacs connected by bike paths, thus making the bicycle a swifter vehicle than the automobile in many instances (as was done in the Village Homes development in Davis, California).
- Speed transit trips by giving transit separate rights of way and the ability to pre-empt traffic signals; instituting an honor system for fares (passengers are on their honor to buy a ticket from a machine or hold a valid pass, enforcement is by random inspection and fine) as is done in Europe and Portland, Oregon; and offering limited stop and express services on major arteries."

—*A Green City Program*

"*Integral Neighborhood.* This 'integral neighborhood' is a very diverse 'mixed use' area. It's integral in the sense that its various functions are closely linked and usefully related one to another. Homes, jobs, schools, recreation, natural features (like the open creek) and agricultural features (like the many gardens and fruit trees) make the neighborhood a kind of village in the city. The basics of a lively community are all here—and so is the culture of the city and all its special contributions that require a larger population base: nearby downtown, schools of higher learning, theaters, research centers, hospitals. . . . In the integral neighborhood there is some animal raising, almost complete recycling, solar and wind energy harnessed, fuel energy conserved with insulation and non-auto transportation. Some professional offices and arts, crafts and trades work spaces fit well here. There's opportunity in nooks and crannies and on specially designed rooftops for childcare, play areas, cafes, places to simply sit and watch the views, people, birds, creek. . . ."

—Ecocity Berkeley

WATER

I T'S TRUE THAT they have 365 faces a year," remarks my friend Debra as we drive by a lovely little lake in early October. Dawn is just beginning to color the sky, and is mirrored in the water's still surface. Tendrils of mist rise from the lake, whose temperature is warmer than that of the air on this chilly morning. A little island clothed in pine trees stands in dark silhouette against the water's luminosity, which joins that of the horizon. Just the sight of water can be soul soothing. An hour later, stroking through the bleachy blue waters of a municipal pool, I'm thinking, water is so smooth, marveling at how almost not-there it feels against the skin, how kindly it buoys one up, how sparkling and yielding it is. About two-thirds of our weight is water. It's the basic constituent of all our tissues, of the fluids that course through our circulatory and lymphatic systems, of tears running down our cheeks, of saliva, mucus, urine, and amniotic fluid, all the waters of bodily life.

Water is so everywhere-present, and such a commonplace, that its gifts seem almost limitless. Water is the only substance on Earth that exists in all three physical states—solid, liquid, and gaseous. Make it hot enough and water turns to steam, a gas with enormous expansive force; make it cold enough and it crystallizes, becoming snow or ice—water you can walk on. It moves gracefully from state to state. Water's physical properties are extraordinary, and its relation to Earth is integral—the biosphere runs on water, sunlight, and minerals. All these ingredients are present on the other planets of our solar system, but the temperatures of those planets are too extreme, too hot or too cold, to allow liquid water in its solvent, circulatory form. Flowing water is a presence unique to Earth among all the planets in our solar system.

On Earth there is so much of it in fact that 71 percent of the planet's surface is ocean. Because of water's extraordinary capacity to absorb and release heat slowly, sizable bodies of water, primarily the oceans, temper climate, maintaining a range of temperatures congenial to life. This tempering is a gift of elemental grace. Ninety-seven percent of the Earth's water is in the salt seas. Most of the fresh water is locked up in the polar ice caps. Thus, less than a percent of the water on Earth is available as fresh water.

Gravity eases it down, solar heat drives it up: precipitation, evaporation, and evapotranspiration continually cycle water through the atmosphere and biosphere. From the deep rolling immensity of the oceans, from the surfaces of lakes and streams, the air takes up water; clouds are formed; dreamy, delicate scarves, massive thunderheads ominous with precipitation—numberless, shifting forms. Some rain clouds slake the Earth gently; others lash with downpours. Fog may simply accumulate in a forest's foliage—like the needles of a redwood tree—and drip quietly to the duff below.

That water always flows downhill is axiomatic; the rate at which it does is determined by the condition of the land on which precipitation falls. Vegetation buffers the impact of rainfall, increases the porosity of the soil and its absorption of water. Roots and other organic matter invite water downward. Rain or snowmelt may percolate through the soil into the Earth's crust. Bodies holding this groundwater are called aquifers and are the source of wells. The level of the groundwater is called the water table.

Precipitation landing fast and furiously—in volume surpassing the surface's capacity to soak it up—qualifies as runoff. When the land's ability to slow runoff is diminished by being paved, logged, strip-mined, overgrazed, or too heavily farmed, erosion and flooding result.

Land use upstream conditions water's way back to the sea, and the fate of all the ecosystems en route. For instance, careless logging upslope speeds runoff, which scours out and silts up the gravel beds where salmon spawn, destroying fisheries. Paving paradise and putting up a parking lot (a lot of lots, actually) seal the surface, giving water nowhere to go but to overload city sewer systems after every big rain.

In undamaged watersheds, precipitation cycles back to the sky by two pathways: Water soaking into the soil, dissolving minerals, is taken up by the roots of plants, moves up their stems by capillary attraction, and out through pores in leaf surfaces to evaporate into the stuff of clouds—evapotranspiration, it's called. It makes for an intimate mutuality between forests and rain. Precipitation falling on steep highlands trickles downhill to join rivulets which become creeks which feed streams which join rivers which flow madly or lazily to the sea. And from the surface of the sea, at the touch of the sun, clouds of water vapor will rise to begin the cycle again, water moving everywhere in the biosphere, the definitive element of earthly life.

Because of our fundamental need for water, human settlements are often found near rivers, on floodplains, and at harbors. As a result of land degradation upstream, floodplains are increasingly chancy places to dwell. The floods that inevitably will inundate them naturally instill a primordial fear. Perhaps that is why the myth of a world flood is a part of many cultures' telling of their distant past. Water has its ominous qualities too. Flash floods or an inescapable drowning are the stuff of nightmares—first the water reaches your windowsills, and then your roofline. What becomes of you, clinging to the roof? Not a deluge, but a gradual insidious flooding may be the result of global warming, which seems to be melting the polar ice caps and causing a slight rise in sea levels. With a rise of only a few inches, coastal settlements may be encroached upon and low-lying islands entirely submerged, catastrophically leading to the displacement of millions of humans. Rising sea levels also could cause the salination of low-lying water supplies.

Trying to make water conform to human will is as ancient a practice as the now-vanished, irrigation-based civilizations of the Indus, Tigris, and Euphrates rivers. Ultimately water will thwart large-scale attempts to store it and move it. Canals and reservoirs inevitably impound not only water, but also silt from upstream. Downstream, fields and fisheries may suffer from being deprived of the nutrients formerly deposited, or borne out to estuaries and deltas by annual floods. Surfaces of reservoirs lose vast quantities of water to evaporation, dissipating the very resource they purport to manage. The long-term futility—and immediate destruction of cropland, indigenous cultures, and riparian habitat—entailed by dams hasn't yet fully registered on federal governmental and international lending agencies. Ever-larger damming and water diversion schemes are still being proposed all over the world.

Water in the shape of rivers gives us images of endless change and flow, ever same and ever changing; some rivers placid, some raging, some broad and muddy, some rocky and hissing, pathways back to the headwaters for salmon and shad, sturgeon and smelt. River damming and water diversion, whether the purpose be for irrigated agriculture or hydropower generation, have devastating consequences, whatever the short-term benefits provided to human populations. Once a wild river is dammed, its wildness and the biodiversity it fosters are killed forever. The United States has sacrificed too many of its free-flowing torrents to dams and reservoirs. Taxpayers effectively subsidize the price of water and power from these dams by picking up the tab for their construction. Not only are rivers silenced and impounded, but the real value of water—particularly in the arid American West—is virtually unknown, which leads to a crazy wastefulness in agricultural, residential, and industrial use of this scarce resource.

Increasingly, agriculture and/or expanding human settlements draw on groundwater supplies from ever-deeper wells or watercourses in ever-receding water tables. Three-quarters of the world's water withdrawal is applied to agriculture; of that, three-quarters is lost to local reuse. If the withdrawal exceeds the aquifer's rate of recharge, what is really going on is water mining, and the wells will eventually run dry, as will wells sunk into underground reservoirs of "fossil" water. As a result of excessive water withdrawals upslope, rivers can dry up and lake levels fall. Over time, irrigated agriculture may salt the Earth (about a third of irrigated land worldwide is now salinized), or waterlog it (a condition affecting about 10 percent of Earth's irrigated lands), and ultimately destroy the soil's productivity. These are patently unsustainable practices.

The human ambiguity toward water in the landscape has also cost the Earth some of its most essential ecosystems—those communities of life found in wetlands. Flooded all or part of the year, wetlands may be the soggy, fecund edge between a lake or pond and the surrounding land; or swamps and sloughs, places where the surface of the land is close enough to the surface of the groundwater to be boggy; or coastal estuaries, river mouths where salt and fresh water mingle, supporting a rich larval stew of aquatic life—fish, crustaceans, insects, wading birds, and other waterfowl. Wetlands are among the most biologically productive habitats on Earth. Coastal and riverine wetlands also gentle the impacts of floodtides or floods on shores and banks, slowing surges and waves with thickets of vegetation—from marsh grasses to mangroves.

Either from an aesthetic desire to draw close to the shore, or from a suspicion of mucky, quaking places, humans have dredged and filled and channeled and diked the planet's wetlands since before Rome rose upon the swampy reaches of the Tiber. Contemporary developers claim that wetlands can be created, and offer to "build" wetlands over here to replace wetlands compromised over there. Because some of the most crucial life-forms in any ecosystem may be tiny, as insects, or even microscopic, and plainly impossible to inventory and cultivate, this claim is essentially false, however well intentioned.

One of the truths most clearly expounded by watercourses is that geometric human political boundaries have an arbitrary quality, cleaving right through watersheds, lake basins, and river systems. These territorial divisions act to the detriment of the waters, reinforcing the misconception of a separate, autonomous existence in persons and peoples: the delusion that it is a state's or nation's prerogative to deforest its uplands; or to dispose of contaminants in, or to overdraw on, its river systems, regardless of the effects on all the lives downstream, and throughout the body of water where they are bound, be it an ocean or an inland sea.

Pervasive elements like air and water are vulnerable to workings of the tragedy of the commons. Given the present values of our civilization, any individual's, or corporation's, state's, or nation's exploitation of a common resource, such as a river system or the ocean, whether it be as source of food (as in fish) or fuel (as in offshore oil drilling), yields an obvious immediate benefit, while the detriments (depletion and pollution) are borne by the entire community. Patterns of responsibility, of risk and benefit, of cause and effect, are confused.

However, in our era the intensity of pollution and overfishing have become so severe that nations are beginning to make earnest efforts to regulate human use of great commons like the ocean floor, the Mediterranean basin, and the Great Lakes. Extraordinary feats of interstate and international cooperation are required. One state or nation's defiance of an accord can negate a complex and painstakingly achieved compromise. The litigious competition among the states in the Colorado River Basin, the refusal of the United States to join in the Law-of-the-Sea compact, a few nations' refusal to join in a moratorium on whaling, for instance, suggest the contention surrounding such continental and global ecological management problems.

If the ecology movement has a global totem, it is probably the whales. The seas are zones of natural mystery. The enormity of whales, their dwelling in the deeps, their songs, their society, their grace, and their possible great intelligence haunt us with the thought that we may not be alone on this planet in having consciousness. A consciousness shaped over millions more years than ours has been, and in a different element, may, although ultimately unknowable, merit some deference and should command some protection. Sadly, after decades of struggle to put an end to whaling and to preserve the few percent of great whales remaining, outlaw nations still refuse to agree to a moratorium on whale killing.

The plight of the whale epitomizes the modern dilemma, the rift of mind that underlies the ecological crisis. To some humans, whales are just a resource, a form of protein, like cattle. They can be eaten, so why should they not? Why shouldn't the investment in factory ships be amortized, even if the "resource" is annihilated in the process? People have to work, and people have to eat. Meat is meat. It is a material view, and we live in a material world. Although leviathans have captured the imagination of school children and adults all over the planet, the apparent failure of this global outpouring of concern to secure the whale's evolutionary destiny is disheartening. What do we have to do to make it come out differently?

*   *   *

It was long thought that running water purified itself every twenty or so feet, a thought which may have eased the conscience of many a lone prospector pissing into a stream. Extrapolating, wrongly, from that principle had us, until all too recently, flushing entire cities' domestic and industrial wastes into river systems, expecting rivers and lakes magically to purify enormous quantities of biological and chemical waste. The sad consequence was that in the sixties we were treated to the spectacle of petrochemically contaminated rivers catching fire and nutrient-laden lakes suffocating from the boom and bust of algae growth. Control of point sources of water pollution and removal of certain chemicals from the household waste water stream (that is, phosphates in detergent) along with more thorough central treatment of human sewage (which, although an improvement, is at best an energy-intensive and chemically dependent approach to the sewage problem) have remedied some of the most obvious forms of water pollution.

Industrial processes have a fierce thirst, and often a need for water as coolant. (Industry accounts for 11 percent of all water use in the United States.) Locating large manufacturing complexes near watercourses and water bodies has led to recent nightmarish disasters like the 1986 Sandoz chemical plant's accidental poisoning of the Rhine River in Germany. Ultimately only a complete revamping of manufacturing processes that generate toxic by-products and dispose of those by-products in the water or air can provide adequate protection for these essential elements. Accidents will always happen, and polluted rivers and aquifers are impossible to clean up completely. Rather than hoping to contain pollutants, we must not generate them to begin with.

Industry's abuse of water is easy to decry and must be dealt with, but it is really just a writ-large manifestation of our culture's taking this element for granted. Most of us in the developed world do.

One afternoon I was doing a little bit of manual labor around my place. It started with my noticing how hot and dry the wind was for late September, and thinking it might be well to sprinkle the garden. I walked over to the house and turned on the faucet. The pump kicked in and started pushing water up from the 220-foot depth of the well. Instantly a glittering spray of droplets arced out over the round plot of vegetables and flowers.

Then I turned my attention to something else: cleaning up a pile of waste lumber near my work space. I wheeled my garden cart over to the jumbled scraps of siding and two-by-fours and loaded it up to add these new pieces to the main pile of construction refuse, all of it eventually bound for the landfill. As I did that I thought about women in villages throughout the Third World having to walk for miles in search of even tiny remnants of brush, wood, or a few branches of desolated trees, for anything to fuel a cooking fire. I thought of how the relentless outward and upward quest for wood, whether it be for the meager needs of the world's marginal households or for the developed world's wants for pulp and lumber and exotic woods, is now deforesting the watersheds of the world. In light of this deforestation, and its effects not just on the land but also on the water and climate, it is strange to be part of a culture in which excess and expedience encourage me to bury wood.

Stranger still is to put turds in potable water, for that is essentially what flush toilets do. Just as they trek for firewood, Third World women and children trek for miles to procure a few gallons of water for their families' needs, returning bowed by the liquid weight. Elsewhere, in cities in the developing world, whole neighborhoods may depend on a single spigot to supply their water. The water it delivers may be pathogenic. In much of the world, human and other wastes still go into open sewers. Nearly two billion people in the developing world lack access to safe drinking water. About seventy thousand people, most of them children, die daily of diseases contracted by drinking contaminated water. Meanwhile, here in the United States, our average daily water use is as much as twenty times that of Third World peoples. In light of global conditions, our society's carelessness with water is strange indeed.

This carelessness can be corrected by awareness and responsibility. Food and water are our most basic needs, and Earth knows how to meet them. Watching Earth attentively, which can teach us how to meet those needs gracefully, reveals a system of relationships and processes. Design strategies learned through the observation of natural systems mean, potentially, that human culture and human settlements could become more sustainable and aesthetic while becoming less destructive of the environments that support them.

Pioneers of biological design, like the aquaculturists of the New Alchemy Institute and the design genius John Todd, of Ocean Arks International, are devising ways to cultivate fish in self-contained greenhouse systems. By incorporating a diverse mix of plant, reptile, fish, and insect species, all of which interact as they would in an ecosystem (albeit in patterns simpler than the inhabitants of natural ponds and rivers would), the Alchemists have created stable food-producing systems whose primary inputs are sunlight and human care and intelligence rather than the higher-on-the-food-chain supplements customary in commercial aquaculture.

John Todd and colleagues are perfecting artificial rivers for biological treatment of sewage, running the effluent through series of tanks containing different combinations of aquatic flora and fauna, which batten on different biological and chemical contents of the sewage, converting "waste" to new life. All waste is treasure, say the Chinese, and the Ocean Arks approach to sewage treatment seems premised on that maxim. The idea that human wastes can be used to build the soil is nothing new. Sophisticated ways to do this that protect water quality and cut the health risks to humans consuming the produce grown from such soil represent a real advance. Meticulous stewardship of all the elements, particularly clean water, is essential to the survival not just of our species, but of the whole Earth.

The hopes for arriving at a saner, wiser relationship to the water resource needn't all depend on ingenious biological design, though. As with energy, conservation begins with commonsense awareness, the discipline of turning off the tap while you're brushing your teeth, for instance, or installing water-reducing showerheads, or putting a weighted plastic jug in the tank to reduce the amount of water flushed down the toilet. There are dozens of household practices that can affirm one's personal consciousness that water is precious, the stuff of life. On a broader scale, it is possible to redesign mining and industrial practices to minimize water use, and methods for recycling polluted water exist, and will surely be employed as the real value of water becomes known, both through rational pricing mechanisms and public education.

Careful stewardship implies a fundamental respect. The tradition of regarding the Earth as a mother suggests the correct attitude: not presuming on the Earth's great generosity, but to become quite sparing in our use of her blessings, and grateful for the very fact of our being, which, like flowing water in the solar system, is a rare gift.

Goodbye to a River
JOHN GRAVES
1960, Alfred A. Knopf

In October 1957, John Graves took his last canoe trip on a stretch of the Brazos River near where he grew up in east Texas. He set out for a three-week adventure with a dachshund pup he called "the passenger," to see for a final time a stretch of river that would soon be dammed—lost forever to the eternal hydraulic progress of the American West. He carried a rucksack of simple food, and a rod and gun to complete his dinners, eating from the river as he had as a boy, depending on it for sustenance made holy by the act of taking.

From this humble adventure, notebook in hand, Graves crafted a work of art, as instructional as it is beautiful, for he understood with uncanny clarity the history of the Brazos country. On every looping bend, at the mouth of each little feeder creek, Graves drew a story mostly from the days of the Indian wars, when Comanches still ruled the Brazos on horseback and routinely savaged the onrushing settlers and were savaged in return. The stories are so persistent, so vividly told, that by the end of the book one suspects that the dams which came to drown the pecan bottoms of this lazy Texas river were more necessary for rinsing out the bloody soil of the countryside and the bitter racial memory that went with it than they ever could be for hydropower or flood control.

In 1957 it was hardly fashionable to write diatribes against dams, and this book is no ringing conservationist elegy. The excerpt contains most of what Graves has to say about the loss of his childhood river.

From the distant vantage of 1990, the remarkable fact about this book is not its eulogizing of another lovely, whole place lost to idiot progress, but the rare feat of simply *knowing a place* so well with no pretence of knowing it scientifically. Practically no American writer now knows any place as well as Graves knew his stretch of the Brazos, or can tell the story of a place with such warm engagement and candor—and that alone makes this book remarkably well worth reading.

—DONALD SNOW

"In a region like the Southwest, scorched to begin with, alternating between floods and droughts, its absorbent cities quadrupling their censuses every few years, electrical power and flood control and moisture conservation and water skiing are praiseworthy projects. . . .

"But if you are built like me, neither the certainty of change, nor the need for it, nor any wry philosophy will keep you from feeling a certain enraged awe when you hear that a river you've known always, and that all men of that place have always known back into the red dawn of men, will shortly not exist. . . . When someone official dreams up a dam, it generally goes in. Dams are ipso facto good all by themselves, like mothers and flags. Maybe you save a Dinosaur Monument from time to time, but in-between such salvations you lose ten Brazoses. . . ."

Cadillac Desert: The American West and Its Disappearing Water
MARC P. REISNER
1986, Penguin Books

This massive piece of work is a superbly entertaining and enlightening history of water development in western United States. Focusing on major projects with intriguing detail, Reisner offers an account of resource use that soars in dramatic appeal and is read by people on opposite sides of the issue. "Pretty accurate," a longtime Bureau of Reclamation engineer told me of the cutting chapter on Bureau Commissioner Floyd Dominy.

Reisner has created a book that not only is a landmark in the story of American resources, but is sure to be a classic in the very history of the West. He has finally pushed water into the mainstream of western historical thought, where popular literature is otherwise stuck on the frontier, the mining era, and the boosterism that has produced shelves of books avoiding troubling truths of our western heritage.

I quickly learned to accept Reisner's boiled-down personal descriptions ("an egomaniacal small-time construction tycoon named Henry J. Kaiser"), requiring trust in the author's judgment, which I found justified by his exhaustive research.

This is not a comprehensive account of water development and is not the story of river conservation. Lacking that element except in a few cases, Reisner's coverage of politics, personalities of dam builders, social-financial inequities, and problems resulting from dams does not represent full coverage of the subject. But one book cannot do everything, and this one does enough.

Because it is their story, *Cadillac Desert* is more than important to every westerner, and is likewise important to other Americans, because—as Reisner points out—they have paid for the growth of the West.

—TIM PALMER

"The Bureau was strikingly candid about the dismal economics of irrigation in the upper basin. 'The [upper basin] farmers can't pay a dime, not one dime,' lamented the Bureau's chief of hydrology, C. B. Jacobsen, to a Congressional committee. And as if to demonstrate how far Congress had come in accepting the subsidization of an entire region, Jacobsen's words fell on sympathetic ears. Western members, even those whose districts were well outside the basin, lined up to support the bill—perhaps because they expected their own uneconomical projects to be supported in return. For the first time, a majority of eastern members seemed indifferent, neutral, or even sympathetic—perhaps because *they* had Corps of Engineers projects they wanted built which might require the western members' support. Even the Eisenhower administration decided to give the Colorado River Storage Project lukewarm support, though it violated every conservative principle Ike had ever espoused."

"The engineers who staffed the Reclamation Service tended to view themselves as a godlike class performing hydrologic miracles for grateful simpletons who were content to sit in the desert and raise fruit. About soil science, agricultural economics, or drainage they sometimes knew less than the farmers whom they regarded with indulgent contempt. As a result, some of the early projects were to become painful embarrassments, and expensive ones. The soil turned out to be demineralized, alkaline, boron-poisoned; drainage was so poor the irrigation water turned fields into saline swamps; markets for the crops didn't exist; expensive projects with heavy repayment obligations were built in regions where only low-value crops could be grown. In the Bureau of Reclamation's quasi-official history, *Water for the West*, Michael Robinson (the son-in-law of a Commissioner of Reclamation) discreetly admits all of this: 'Initially, little consideration was given to the hard realities of irrigated agriculture. Neither aid nor direction was given to settlers in carrying out the difficult and costly work of clearing and leveling the land, digging irrigation ditches, building roads and houses, and transporting crops to remote markets. . . .' "

"In the West, many soils are classified as saline or alkaline. Irrigation water percolates through them, then returns to the river. It is diverted downstream, used again, and returned to the river. On rivers like the Colorado and the Platte, the same water may be used eighteen times over. It also spends a good deal of its time in reservoirs which, in desert country, may lose eight to twelve feet off their surface to the sun every year. The process continues—salts are picked up, fresh water evaporates, more salts are picked up, more fresh water evaporates. The hydrologist Arthur Pillsbury, writing in *Scientific American* in July of 1981, estimated that of the 120 million acre-feet of water applied to irrigate American crops the previous year, ninety million acre-feet were lost to evaporation and transpiration by plants. The remaining thirty million acre-feet contained virtually all of the salts."

The Great Lakes: An Environmental Atlas and Resource Book
JOINTLY PRODUCED BY ENVIRONMENT CANADA,
U.S. ENVIRONMENTAL PROTECTION AGENCY,
BROCK UNIVERSITY, AND NORTHWESTERN UNIVERSITY
1987, U.S. Environmental Protection Agency
and Environment Canada

The Late, Great Lakes: An Environmental History
WILLIAM ASHWORTH
1986, Alfred A. Knopf

Reflections in a Tarnished Mirror: The Use and Abuse of the Great Lakes
TOM KUCHENBERG
1978, Golden Glow Publishing

Standing on a Lake Michigan beach, facing west as a storm boils across from Wisconsin—churning the waters into a chaos, engaging in shadow play with the contours of the offshore islands—watching such a spectacle take place in the middle of a continent makes you aware that *lake* is an inadequate term to describe such horizonless immensity. The Great Lakes macroregion is vast, scenic in some of its reaches, an industrial sewer in many others. You think twice about eating fish caught in the Great Lakes, and a lot of people who live around them die of cancer.

These "sweetwater seas" make America's fourth coast, and their watershed is just so vast, says Lee Botts, a longtime Great Lakes activist, that "people can't relate to the whole system."

For that reason, perhaps, there is no single book that quite does it justice. The EPA and Environment Canada's *The Great Lakes* is a good first book, a current, straight-down-the-middle atlas providing basic geologic, hydrologic, socioeconomic, and environmental quality data in text, charts, and maps. A joint production of U.S. and Canadian government agencies, *The Great Lakes* is available free from Great Lakes National Program Office, U.S. Environmental Protection Agency, 230 South Dearborn Street, Chicago, Illinois 60604.

William Ashworth's *The Late, Great Lakes* is a lively introduction to these inland seas and the saga of the pollution visited upon them by burgeoning onshore development, but it was based on a rather too brief inspection tour and suffers from inadequate fact checking. Nevertheless, says Botts, "There really isn't a good substitute."

Reflections in a Tarnished Mirror, by Tom Kuchenberg with photographs by Jim Legault, is a documentary, in book form, of the development and change of the Great Lakes sport, commercial, and Indian fisheries. The book recounts, through candid interviews and true-grit images, their boom and bust cycles, the fisheries' diminishment by onshore pollution (with logging wastes preceding toxic chemicals as a culprit), and their ecological revolutions as a result of the deliberate and inadvertent introduction to the lakes' ecosystems of exotic fauna such as the coho salmon, the alewife, and the lamprey.

The fate of the Great Lakes is the fate of us all. "There isn't any such thing as *just* a Great Lakes problem," says Lee Botts. "If it affects the Great Lakes system, you've got to assume that it's a global problem." But when there is a Great Lakes solution, as in the lowering of phosphorus loadings in the lakes, she says, it represents "unparalleled success in international environmental management," and hope for us all.

—STEPHANIE MILLS

"Now the Great Lakes basin is home to more than one-tenth of the population of the United States and one-quarter of the people of Canada. Some of the world's largest concentrations of industrial capacity are located in the Great Lakes region. Nearly 25 percent of the total Canadian agricultural production and seven percent of the American production are located in the basin. The United States considers the Great Lakes a fourth seacoast and the Great Lakes region is a dominant factor in the Canadian industrial economy.

"The magnitude of the Great Lakes water system is difficult to appreciate, even for those who live within the basin. As a whole, the lakes contain about 23,000 km^3 (5,500 cu. mi.) of water covering a total area of 244,000 km^2 (94,000 sq. mi.). The Great Lakes are the largest system of fresh, surface water on earth, containing roughly 18 percent of the world supply. Only the polar ice caps contain more fresh water."

"*Fish Advisories*
"The state and provincial governments surrounding the Great Lakes have issued advisories for people consuming fish caught in the lakes. These advisories suggest that consumption of certain species and sizes of fish should be avoided or reduced due to toxic chemicals present in the fish. These chemicals can cause a number of human health problems ranging from cancer to birth defects and neurological disorders.

"As a result of uncertainty in the scientific community about the toxicity to humans of these chemicals, the jurisdictions surrounding the lakes vary in the advice they provide. However, in all cases, following the advisories will reduce (but not necessarily eliminate) the exposure and, therefore, the risk of suffering adverse effects. High-risk groups such as pregnant women, nursing mothers and preteen children are advised to pay close attention to the advisories."

—*The Great Lakes*

"The Big Cut was an era of extravagance. There was a road from Saginaw to Flint literally paved with clear pine, and a mountain of sawdust at Cheboygan 40 feet high and nearly a mile around. There were clearcuts that covered entire townships, the moraines and kames and drumlins (elongated or oval hills) left by the continental glacier sticking up obscenely through the naked soil like the ribs of a shaved dog. Cedars were shriven for matchsticks; hemlocks were felled, stripped of their bark for its tannin, and left to rot where they lay like so many buffalo."

—The Late, Great Lakes

"The lake sturgeon was the second fish to go into decline. Though pushed to the edge, it has not followed the Atlantic salmon into extinction. It is a magnificent creature, reaching lengths of 8 feet and weights of 300 pounds. Potentially, it has an incredible life span that may reach 150 years. Ironically this life span may have been its chief difficulty, for it does not reach spawning age for a leisurely 14 to 23 years and does not spawn in every year. Its near disappearance presents the strongest case for overfishing. It was first regarded as a nuisance because of the low market value and the destruction of nets the giant caused when trapped. So plentiful and lightly regarded was the fish, that it is reported to have been used as fuel under the boilers of early Great Lakes steamboats. The deliberate destruction of the fish must have been considerable. Compounding the problem was the discovery of a number of uses for the sturgeon. This led to an upsurge in harvesting that in Lake Ontario led to a catch of 581,000 pounds in 1890. The fish thereupon became increasingly rare, reaching levels of less than 10,000 pounds by the early 1920's."

"The destruction of the commercial fishing industry that was wrought by the lamprey and the invading species was significant. All across the upper lakes, marginal operations folded and for those families to whom fishing had become a way of life, it was a time of getting along as well as one could. Members took other jobs or attempted to struggle along with either lower value species or reduced incomes. By the mid-sixties, however, the fishermen looked with great hope to the increasingly successful lamprey control programs and the subsequent restocking of Lake Superior with trout. This program was soon scheduled for Lakes Michigan and Huron. Fish management officials had generally turned a sympathetic ear to the needs of the commercial fisherman. Fishing was a valuable source of food, an important factor in the economy of numerous lakeside communities, an honored way of life, and the prime source of what little knowledge management personnel possessed on the nature of the ecosystem and the environment of the Great Lakes."

—Reflections in a Tarnished Mirror

We All Live Downstream: A Guide to Waste Treatment That Stops Water Pollution
PAT COSTNER WITH HOLLY GETTINGS AND GLENNA BOOTH
1986, National Water Center*

Among city dwellers who moved to rural areas in the 1970s, one key area of learning was what to do with our own shit and waste water. I got so interested in the subject that I designed an above-ground privy designed to decompose human wastes and eventually wrote a book about the subject (*The Toilet Papers,* 1978, Capra Press, out of print), which will be reissued soon.

Perhaps the best book still in print on no-flush and low-flush toilets and related issues is *We All Live Downstream.* The book offers a comprehensive overview of water-conserving ethics and technologies. While its virtue is a broadbrush overview, its weakness is that it ignores the last fifteen years of experience and experimentation with these systems. Proprietary systems are illustrated and described based on manufacturers' literature, but there is no evaluation by users. (The Farallones system I designed, for example, was mod-

*National Water Center, P. O. Box 548, Eureka Springs, Arkansas 72632

ified over the years based on feedback from owner-builders.) There is no discussion of building and sanitation code issues, despite fifteen years of battling local health and building departments that refuse to permit most non-water-based disposal systems.

Indeed the most interesting part of the book is a discussion of the political campaign against a central sewage plant for Eureka Springs, Arkansas, out of which the Water Center was organized. In spite of a six-year campaign by the eco-activists for the city to consider alternatives, a three-million-dollar central sewage plant was built, and there was no funding for a decentralized compost-toilet-based system.

—SIM VAN DER RYN

"*Low Flush Toilets*

"To achieve an even greater saving of water, a low-flush toilet can be installed. Since the toilet uses almost 40 per cent of the average household water and produces the blackwater which is the most difficult to repurify, this is a rewarding target for water conservation.

"Most of the low-flush toilets require no special plumbing alterations and work like a regular flush toilet. They use less water because of design changes in the bowl shape, trap pitch and water column height. They are easy to install and results are immediate. A few models are detailed here. Information is taken from the manufacturer's product literature."

"Consider the milk cooler in your grocery store. It may hold as many as 300 one-gallon jugs of milk. Fill all 300 jugs with water; stack them in your living room. You now have a vivid example of the amount of clean, drinkable water that the average U.S. family of four uses every day.

"About 95 per cent of this water, 285 gallons, ends up as sewage. 120 gallons is flushed down the toilet and the remaining 165 gallons go down household drains carrying shampoo, detergent, cleanser, food scraps and dirt.

"Obviously a dry toilet saves more water than any other water-conserving device, since it reduces water use by about 40 per cent. With faucet aerators, flow-restricting showerheads, and other shelf items from the local hardware store, that typical family's stack of 285 jugs of water-to-become-sewage can be reduced to only 60 gallons of greywater.

"Water-saving devices are generally inexpensive and available at hardware stores and plumbing supply houses. Most can be installed by the owner in less than 30 minutes.

"A standard bathroom and kitchen can be retrofitted for water conservation for as little as $24. Savings in water heating expenses alone will return the full investment within three months with a conventional water heater. For a tankless, on-demand heater, the payback is less than one month."

Love Canal: My Story
LOIS MARIE GIBBS, AS TOLD TO MURRAY LEVINE
1982, Grove Press

It's hard to describe my relationship with Lois Gibbs over the years; I never met her, but I get a warm feeling just from knowing she's still there, toilet training corporate America. From a "dumb housewife" twelve years ago, she's somehow been transformed by the buried hazardous wastes of Love Canal, the resulting injuries to her son, and the similar plight of others into the single most effective leader of the American antipollution effort. We will always owe Lois and the friends she led for radical improvements in social attitudes, law, and our environment. This 1982 biography describes the emergence of one of the most successful revolutionaries of our century.

—CAROL VAN STRUM

"I was stunned that the school board had allowed a school to be built on such a location. Even today, it doesn't seem possible that, knowing there were dangerous chemicals buried there, someone could put up a *school* on the site. The 99th Street School had over 400 children that year, one of its lowest annual enrollments."

"As I continued going door-to-door, I heard more. The more I heard, the more frightened I became. This problem involved much more than the 99th Street School. The entire community seemed to be sick! Then I remembered my own neighbors. One who lived on the left of my husband and me was suffering from severe migraines and had been hospitalized three or four times that year. Her daughter had kidney problems and bleeding. A woman on the other side of us had gastrointestinal problems. A man in the next house down was dying of lung cancer and he didn't even work in industry. The man across the street had just had lung surgery. I thought about Michael; maybe there *was* more to it than just the school. I didn't understand how chemicals could get all the way over to 101st Street from 99th; but the more I thought about it, the more frightened I became—for my family and for the whole neighborhood."

"Love Canal is not over. The families will suffer from Love Canal the rest of their lives. If the Revitalization Committee has its way, they will resell most of the homes to new, innocent victims. Five or ten years from now, you will probably hear the cries from people at Love Canal again. The residents of Love Canal learned a lesson; I'm not sure that government and industry have. It will be up to us, as citizens, to tell them forcefully they can't cover over Love Canals. All our lives are at stake."

The Poisoned Well
Sierra Club Legal Defense Fund
1989, Island Press

That old saying "what goes around comes around" is especially true for water. It moves in a cycle through our atmosphere, earth, oceans, lakes, rivers, and ourselves. Our bodies are more water than anything else. We depend on it for life. Yet every day we hear about another community where the water now brings sickness instead of health. And some of us haven't simply heard of it, we've lived it.

The Poisoned Well focuses on that portion of the water's cycle which is least visible but upon which Americans are most dependent. Ninety-six percent of all the available fresh water in the United States is groundwater. This supply is threatened by chemical contaminants from our history of careless disposal.

Written by members of the Sierra Club Legal Defense Fund, the book is an excellent resource for those of you ready to fight for your right to clean drinking water. Don't worry if you're not sure where your water comes from, much less whether it's threatened by contamination. *The Poisoned Well* starts with all the basics and quickly moves into the legal and administrative tools available to citizens at the federal, state, and local level. It's arranged by the types of contamination sources, so you don't have to wade through a long discussion on a particular law to find which sections apply, for example, to your community's leaking underground storage tank.

One of the most commendable aspects of *The Poisoned Well* is the authors' continuous encouragement to utilize the law *along with* many other tactics available to grass roots citizens' groups. The book gives us some good examples of how citizens' groups that combine their local talents, smarts, and imagination have won big victories and forced their governments and industries to follow the law of the land.

—LOIS MARIE GIBBS WITH
PAMELA K. STONE

"All of the programs discussed in this book are similar in one respect. They provide only a framework for action. The laws themselves do not protect groundwater. It takes aggressive and effective enforcement action to ensure that the law is translated into clean groundwater.

"Many people assume that government agencies alone are responsible for enforcing the law and providing the protection the law envisions; but all too often agencies do not fulfill their role. This inaction may happen for a variety of reasons ranging from budget shortfalls and meager staffs to agency policies or personnel opposed to aggressive enforcement.

"Whatever the reason, when agencies do not do their job, citizens have to step in and do the work themselves. In the end, the laws and programs discussed in this book work only as well as you make them work."

RACHEL CARSON

The mistakes that are made now are made for all time.
                                                    —The Sea Around Us

The balance of nature . . . cannot safely be ignored any more than the law of gravity can be defied with impunity by a man perched on the edge of a cliff.
                                                    —Silent Spring

Silent Spring is a manifesto. Rachel Carson undertook the seemingly impossible challenge of impassioning the public about the subject of pesticides, inciting both a barrage of opposition and a following vital enough to launch the environmental movement. Few books in history have so catalyzed the world's attention, inspiring a wave of government hearings, the banning of DDT, and the eventual creation of the Environmental Protection Agency. What sustains the book as radical is its unflinching determination to raise questions that go to the root, challenging the very foundations of our social, economic, and spiritual relationship to nature. What makes the book so compelling is its voice. Rachel Carson is no ideologue. Her outcry against the reckless contamination and destruction of the Earth rises from a deeply intelligent, unsentimental brand of love and reverence for all life. Nearly three decades since *Silent Spring*'s publication in 1962, we are besieged daily by the consequences of her unheeded prophecies.

Silent Spring dismantles the official platform constructed by science, industry, and government that the introduction of pesticides and other toxic chemicals into the environment is "harmless." On

145

the contrary, Carson discloses the gruesome details of acute poisonings of humans, vegetation, and wildlife resulting from "insect control programs," mounting page after page of unshakable evidence
that the authorities' pronouncement of "no risk" is a myth based on
countless forms of ignorance and deception. Her charge that we
inform ourselves about the dangers of this chemical tinkering with
nature requires educating the average reader about everything from
the makeup of the carbon atom to the contribution of the earth-
worm. Only after explaining the biology of how pesticides make
their way into the marrow of bones and the substance of cells can
Carson unveil the most alarming effects of long-term exposure to the
battery of chemicals now present in our environment: the development of cancer, genetic mutations, and the "menacing shadow" of
the future of sterility and extinction.

Carson's gift as a writer for translating the world of geology,
biology, and chemistry into stirring, evocative prose lures the lay
reader through the otherwise inaccessible gates of the domain of
science. This, in itself, is revolutionary in Carson's view: to enter the
land of the specialist in order to question the decisions enacted, too
often, without public scrutiny. The development of this gift for
language, and the unfolding of Carson's convictions about our
moral obligation to respect the integrity of every living existence, can
be traced in her first three books: *Under the Sea-Wind, The Sea
Around Us,* and *The Edge of the Sea.* These books, the first two of
which she began while employed as a biologist for the Fish and
Wildlife Service, were the outcome of Carson's ". . . absorption in
the mystery and meaning of the sea. . . ." All three works seek to
inspire awe and locate human life on the periphery in order to
undermine our anthropocentric posture toward the natural world.

It is this position against human arrogance that Carson eventually
unleashes in *Silent Spring.* Despite much popular acclaim, an elaborate and expensive campaign was launched to silence *Silent Spring.*
Radio commentaries, lecture tours, and official chemical and agricultural pamphlets all appeared to discredit Carson as a scientist,
and her book as propaganda. Those implicated in her book, those
with a vested economic interest in the research, marketing, production, and distribution of pesticides, rose up to suppress its insistent
message. Paul Brooks, Carson's editor and biographer, writes of this

assault (in *House of Life,* 1972, Houghton Mifflin Company), "The fury with which it was attacked, the attempts to discredit that 'hysterical woman' . . . have, I believe, deeper roots than a simple concern for profits or power on the part of special interest groups. Her opponents must have realized—as was indeed the case—that she was questioning not only the indiscriminate use of poisons but the basic irresponsibility of an industrialized, technological society toward the natural world."

Carson warns us that it is not mastery over nature that is necessary, but the maturity that comes with being capable of controlling ourselves. Instead, we behave as infants, acting as if every living thing and being on Earth were invented for the sole purpose of our gratification. Carson reminds us, both as comfort and as warning, that we are not alone here. We are only one thin strand in the "fabric of life," charged by her to take responsibility for our recently acquired power to destroy all existence. Reading *Silent Spring,* it's as if one begins to hear Carson's voice insisting between the lines—*Look. Your life depends on what I'm telling you. And not only our lives.* We need her vision now as never before.

—LINDA-RUTH BERGER

Silent Spring
1962; 1987, Houghton Mifflin Company

Under the Sea-Wind
1952, Oxford University Press, out of print

The Sea Around Us
1954; 1989, Oxford University Press

"In each of these situations, one turns away to ponder the question: Who has made the decision that sets in motion these chains of poisonings, this ever-widening wave of death that spreads out, like ripples when a pebble is dropped into a still pond? Who has placed in one pan of the scales the leaves that might have been eaten by the beetles and in the other the pitiful heaps of many-hued feathers, the lifeless remains of the birds that fell before the unselective bludgeon of insecticidal poisons? Who has decided—who has the *right* to decide—for the countless legions of people who were not consulted that the supreme value is a world without insects, even though it be also a sterile world ungraced by the curving wing of a bird in flight? The decision is that of the authoritarian temporarily entrusted with power; he has made it during a moment of inattention by millions to whom beauty and the ordered world of nature still have a meaning that is deep and imperative."

—Silent Spring

"It is always the unseen that most deeply stirs our imagination, and so it is with waves. The largest and most awe-inspiring waves of the ocean are invisible; they move on their mysterious courses far down in the hidden depths of the sea, rolling ponderously and unceasingly. For many years it was known that the vessels of Arctic expeditions often became almost trapped and made headway only with difficulty in what was called 'dead water'—now recognized as internal waves at the boundary between a thin surface layer of fresh water and the underlying salt water. . . .

"Now even though mystery still surrounds the causes of these great waves that rise and fall, far below the surface, their ocean-wide occurrence is well established. Down in deep water they toss submarines about, just as their surface counterparts set ships to rolling. They seem to break against the Gulf Stream and other strong currents in a deep-sea version of the dramatic meeting of surface waves and opposing tidal currents. Probably internal waves occur wherever there is a boundary between layers of dissimilar water, just as the waves we see occur at the boundary between air and sea. But these are waves such as never moved at the surface of the ocean. The water masses involved are unthinkably great, some of the waves being as high as 300 feet."

—The Sea Around Us

The Earth Manual: How to Work on Wild Land Without Taming It
MALCOLM MARGOLIN
1975; 1985, Heyday Books

Restoring the Earth
JOHN J. BERGER
1985, Alfred A. Knopf

The first time I came across *The Earth Manual,* in the office of a restoration group on California's North Coast, I forgot what I was supposed to be doing and browsed through it for an hour. I day-dreamed about the projects Malcolm Margolin describes, just as cooks read recipe books. His instructions for controlling erosion, creating ponds, and propagating wild plants transported me to a nearby creek and dressed me in mud boots.

Margolin interweaves these directions with anecdotes about his work with kids and other volunteers in a large San Francisco Bay Area park. And he does it with a sense of humor that veterans of environmentalism will treasure. He cheerfully admits the pleasure that he and his volunteers have felt at felling a tree, and he pokes fun at the sentimentality that some people feel at the thought of planting one. (Directing folks who each plant a hundred trees in three hours probably erodes that sort of mushiness.) Margolin tempers his enthusiasm with the humility of a man who has worked on so many restoration projects that he has made his share of mistakes and knows both the limits of what humans can help nature do and the magic of how humans are themselves transformed in the process.

(For an idea of what this sort of work has accomplished across the continent, try John Berger's *Restoring the Earth.* Berger recounts the transformations wrought by people in cleaning up industrial rivers in New England, reintroducing peregrine falcons to the California coast, building marshes on the Atlantic coast, and planting prairies in the Midwest.)

There's a balance to be struck in restoration work between cod-

149

dling nature and backing off so far that one's efforts have little effect. In *The Earth Manual: How to Work on Wild Land Without Taming It*, Malcolm Margolin walks that line with elegance and inspiration. The book's sense of proportion is even reflected in its cover image, depicting a barely noticeable human-with-shovel, framed by a spreading oak and silhouetted against looming hills.

—SETH ZUCKERMAN

"Wildlife needs food, cover, and water. It needs all these things within a fairly compact area—how compact depends on the size of an animal's 'territory.' If any one of these things is missing, or in short supply, this will limit the number of creatures the land can support.

"The next time you look at a piece of land, chant over and over again, like a mantra,

> Food, cover, water;
> Food, cover, water;
> Food, cover, water.

It will help you see the land as an animal might see it. A thick forest may have plenty of cover but no food. An irrigated farm may have plenty of water and enough food to feast Noah's Ark, but if there is no cover, it will be virtually a wildlife desert."

"To begin at the beginning, drops of rain fall down. Plip, plip, plip. They hit the ground at a speed of about thirty feet a second. If your land is healthy and the raindrops fall onto a thickly carpeted meadow, a wonderful thing happens. It is something you have to see to appreciate fully. The next time it begins to rain, try to forget everything your mother taught you about 'catching your death of cold,' lie down on your belly, nestle your chin into the grass, and get a frog's-eye view of how raindrops fall. You'll see how the raindrops hit the individual blades of grass, causing them to bend down. This bending absorbs the energy of the raindrop, and the raindrop slides gently off the blade of grass, which immediately springs up again, waiting to catch another raindrop."

—*The Earth Manual*

Waste to Wealth: A Business Guide for Community Recycling Enterprises
JON HULS AND NEIL SELDMAN
1985, Institute for Local Self-Reliance*

You want to set up a community recycling program. Or you're studying the technology or economics or business or sociology of recycling. So you get invited to one of those rare lectures given by people who've learned by doing. They share their pragmatic, workable suggestions and steer clear of environmentalist rhetoric. When the lecture is over, you're handed well-organized, typewritten notes, complete with lookup charts.

That's what this book is like.

Waste to Wealth lets you rub elbows with people in the *business* of recycling. After a short introduction that shares rules of thumb for feasibility, the next section outlines a plan for municipal curbside collection of glass, cans, and newspaper from residential neighborhoods. Then, the biggest section of the book gives technical and cost-related data about these entrepreneurial ventures:

- high-grading wastepaper
- manufacturing cellulose-fiber products from wastepaper
- bottle washing for reuse
- processing used oils for fuel
- processing crumb rubber for road construction

The authors deliver an honest appraisal of problems with recycling: market fluctuations and glut, noncompliance, transportation costs. You'll get excellent information on both capital and operation costs, such as how to figure start-up, vehicle, and labor costs and how to establish buy/sell prices.

Books and pamphlets for further reading are listed at the end of each section, complete with ordering information. An appendix lists

* Institute for Local Self-Reliance, 2425 Eighteenth Street, NW, Washington, DC 20009

technical consultants for markets, equipment, education, program planning, and economic development. Another lists manufacturers and suppliers of refuse containers, balers, shredders, scales, collection vehicles, conveyors, separators, and glass crushers.

Waste to Wealth wasn't written to convince consumers and legislators of the benefits of recycling. Instead, it's a no-frills business manual. Nevertheless, this manual of practice, with its reality-based facts and figures, makes an irrefutable case for the feasibility of recycling.

—PATRICIA POORE

"Interestingly, about 75 percent of all wastes generated and disposed are of mixed organic origin. There would appear to be some potential for either burning waste for its energy content or composting it for its soil enhancement qualities. A successful materials recycling program will improve the quality of the remaining waste as a feedstock for either of these recovery approaches."

"*Bottle Washing*
"Washing is normally done automatically. A small system handles less than 200,000 cases per year. Medium sized systems can handle up to 1,000,000 cases per year. The speed with which the bottles are handled depends on their size, soaking requirements (if any), bottle preparation, and special bottle design characteristics.

"Bottle washing functions require a minimum of 10,000 square feet of enclosed or covered space. Also necessary are adequate utility hookups and ventilation, transportation availability, and a loading area large enough to accommodate both incoming and outgoing deliveries simultaneously.

"Processing equipment includes the modular bottle washer, boiler, pallets for storage, and a fork lift or hydraulic hand lift. The intermediate and larger size operations have conveyor belts and distributional case formers, partitions, and packers (the last three items are used to package washed bottles).

"Labor requirements vary depending on the size of the plant and whether it is separate or integrated with a bottle filling operation. In a separate facility labor varies from 6 to 53 persons. In a small facility handling less than 200,000 cases per year each employee may assume several roles. In larger facilities there will be more differentiation in roles. The labor requirements for a small-, medium-, and large-sized facility are given in Table 3-9. Except for management, no special technical training is necessary."

Bioshelters, Ocean Arks, City Farming: Ecology as the Basis of Design
NANCY JACK TODD AND JOHN TODD
1984, Sierra Club Books

My goodness, what could all those items in the title have in common? Well, besides being fashionably eco-righteous subjects, they're all real, full-scale experimental alternative technology projects initiated and nurtured by the authors over the past twenty years or so. From the Todds you get not mere protest, no easy potshots at big corporations, no hypocritical sniping at that amorphous and implacable enemy, technology. Indeed, most of the concepts actualized by the Todds blend the best of high tech with biological principles to achieve their worthy goals. There's no simplistic "all-you-gotta-do-is" stuff though; the projects are pragmatic and they work. Once up and running, they encourage further learning. New ideas are born, inspired and informed by the success—and failure—of earlier work.

This book chronicles the adventure of the Todds and some of their friends. Copious quotes from colleagues and fellow adventurers show the logic and lineage of the ideas while imparting a feeling that we are a "we"—part of a movement just getting under way. There's an air of hope and vitality to it all. In other words, this is practical advice; a proven, credible, and inspiring environmental design primer.

—J. BALDWIN

"*Precept Six—Design Should Be Sustainable Through the Integration of Living Systems*

". . . using the biological world as a model, it has been New Alchemy's intent to, whenever possible, integrate design and function. A solar pond, for example, is an aquaculture unit, a heat storage unit, and low level furnace."

"Some forms of aquaculture or fish farming make particular sense in towns and cities. Fish are most palatable when fresh. The New Alchemy model of solar aquaculture has proved that it is both possible and cost effective to grow fish and shellfish in small spaces, using relatively little water. The translucent, cylindrical solar-algae pond provides a superb habitat for algae-based ecosystems containing cultured fish. Such tanks can be placed in almost any spot that receives direct sun, not the least of which is inside a bioshelter."

SPIRIT

I T GOES BY DOZENS of names, and different peoples see it expressed throughout all life: spirit, something larger and more mysterious than the individual self. As humans, our conception of soul and psyche is largely shaped by the human mind. However, with the image of Gaia, and our deepening understanding of ecology, it seems the life force itself shows intelligence—mind in nature.

Everything connects; we may not all be aware of it, but we do interact with all life; we all bear some measure of responsibility for life's future. In this age of information and spiritual exploration, the knowledge necessary to effect change on behalf of the life that lives through us, and through which we live, has never been readier to hand.

And at this turning point in life's history on Earth, the human variety of consciousness is both a blessing and a curse. It was mind (and opposable thumb), after all, that got us into this mess. And mind's gifts to life—human creativity, community, cultural diversity, learning, love, spiritual aspiration, and art, to name but a few—are what will lead us out.

The species-wide soul crisis precipitated by the global environmental crisis offers us the opportunity to mature; a sadder but wiser species, what we are here for now is: reclaiming, restoring, preserving, protecting, atoning. We have the opportunity to become more real to ourselves now, and to see ourselves in full dimension.

Faith, dogged determination, altruistic actions, believing in your beloved when they've lost belief in themselves, striving to experience

157

oneness with the universe, studying to divine the meaning of sacred texts and sacred syllables, fearless surrender to the infinite, epiphanies on the shore—the human striving for higher consciousness and greater compassion is a limitless wonder. It is a yearning latent in everyone, indomitable despite the privation and alienation into which the human species is wont to plunge.

When work on this book was just beginning, I'd been browsing through a color atlas of Hopi kachina "dolls." (Like a lot of people whose attitudes were ordered—or disordered—in the sixties, I am fascinated by the romance and mystique of this people, an American Indian tribe living a sacred, sustainable lifeway in northeastern Arizona, a remote, dramatic, arid, and highly exacting region.) The mesa-top Hopi pueblos are the oldest continually inhabited settlements on our continent. Despite the present federal threats to their cultural survival, traditional Hopi emissaries have traveled the world in an effort to share their understanding of the nature of balance and to plead for their right to continue as a people. The loss of this, or any culture's earthly wisdom, is grievous to every culture. The global community not only needs new knowledge to get there from here, but also to attend the old ways too.

Reflecting on the significance of the dolls to Hopi culture, I made a cryptic note to myself: "Kachinas and environmental education." In our culture better environmental education is an obvious necessity. But how to educate truly? Learning is a function of a nuturant relationship, and the living world is the greatest teacher. As an example, here is how the kachinas educate the Hopi. The kachinas are not exactly gods, but indwelling tutelary powers in a myriad of plants and animals and phenomena that *are* the world of the Hopi.

"The basic concept of the [kachina] cult," writes Barton Wright in *Hopi Kachinas*, "is that all things in the world have two forms, the visible object and a spirit counterpart, a dualism that balances mass and energy. Kachinas are the spirit essence of everything in the real world. . . ." Listen to the English versions of the names of just a few of the kachinas: Antelope Kachina, Assassin Fly Girl, Badger Kachina, Big Ears Kachina, Blue Corn Girl, Broad-faced Kachina, Butterfly Man, Chasing Star Kachina, Chili Kachina, Cloud Guard Kachina, Cocklebur Kachina, Crow Mother, Cumulus Cloud Kachina Girl, Dung Carrier, Great Horned Owl, Hand Kachina, Hump-backed Flute Player, Lightning Long-haired Kachina, Meteor Kachina, Prickly Pear Kachina, Star Kachina, Sweet Cornmeal Tasting Mudhead, Water Serpent Kachina.

"The Hopi do not worship these kachinas," says Wright, "but rather treat them as friends or partners who are interested in Hopi welfare. Because it is not easy to interact with the kachinas in their insubstantial form, it remains a simple matter to give them shape and personality by impersonating them." During the year the kachinas visit the Hopi and make their presence felt in a ceremonial cycle of dances and songs that are performed by members of sacred societies and attended by the entire community.

"The men who impersonate kachinas and dance in the plazas carve small wooden replicas of their kachina appearance and present these to infants and all ages of females," Wright continues. "This carved and painted figure is called a *tihu* by the Hopi and a kachina 'doll' by others. It is not a doll, a plaything for children, but an effigy or small part of the kachina it represents. . . ." The imagery of the costumes and dolls is never literal but startlingly abstracted, visual intuition of indefinable things, heightened in significance by the symbolic content of every detail, from the colors in which they are painted to the different kinds of feathers that once would have adorned them.

So the kachina *tihus* are tokens of the powers that determine the world, and the kachina ceremonials are occasions of teaching as well as reverencing. Everybody instructs the young, and the message is that one does not live merely for oneself, that one lives for the people, and for the continuity of the way of life. Living in a rigorous environment (which through human agency all environments are becoming) demands discipline. And if this discipline is imposed latterly or externally, it's likely to be an ill fit. So the genius of traditional peoples is that they teach their lifeways continually, and by a variety of means, from cradle songs to pervasive symbolic connotation attached to seemingly commonplace objects, to epic creation tales, to ritual offerings in specific places, to legitimating the authority of all elders in the tribe to teach the correct lifeway, the lifeway appropriate to place. In this the Hopi culture is making perfect use of the human gifts of memory, language, artistic creativity, and humor. Within this integument of culture, the tribe itself is an organism, inseparable from its surround.

However much I may envy the sacred knowledge of a people like the traditional Hopi (although hardly envying their struggle to survive), I can't be Hopi. Rather I have to reflect and act on what movement toward Earth-reverence is attainable in my own culture. Ruminating on this one mid-October morning, I went into town to watch a friend teach a children's aikido class. It was just getting light, the sky full of cottony gray rain clouds, the landscape heading toward winter. Much of the glowing gold and flame of the foliage had been dashed to the ground by recent rains and a snowfall. The new wetness was bringing out the intensity of the remaining colors, blackening the bark of the trees, etching their shape and character against the sky now that the cloak of leaves was ripped away. Cresting over the hill, I saw the sunrise causing the underside of the clouds to blush briefly, a glory over the saturated teal of the bay's waters. There was an epiphany in this, a moment of seeing the magic inherent in an everyday sight of farmland, sky, and shore.

Onward into the ordinary reality of the town on an early Saturday morning, among my fellow citizens out on errands or looking for a little recreation. And out to the gym at the junior college. The class consisted of maybe a dozen children, most wearing the aikido *dogi*, a simple, loose-fitting, white costume, donned and belted in a certain manner that connotes the principle of aikido: nonresistance, a way of finding harmony. There were four adults—my friend teaching, three others helping. The kids ranged in age from six to ten, and there were more boys than girls. It was touching to watch these young ones trying something new, something foreign. Here were kids tasting a discipline, taking the big visible risk of learning through trial and error. They all looked lithe and lean, compact and muscular as puppies, and were pretty much paying attention. Some were learning by observing; others were not quite getting it. It was a world, that class, aquariumlike under the merciless gymnasium lights. Engrossing, but not tender. Unsentimental, and sometimes funny. Looking at those young ones, I ached, knowing that, much as I or any of us might desire it, I can't fix the world enough to forestall the suffering their lives will inevitably entail. Life in her profligacy exacts a huge toll.

Rachel Carson, in *At the Edge of the Sea*, describes different oceanic and estuarian life-forms, creatures whose spawn may number in the millions. And among those millions, perhaps only a few will survive into maturity. Most of them will be eaten. The implacable destiny of the individual organism is first to live and then to die. There's an undeniably tragic edge to this reality. That those aikido kids, indeed all kids, are emerging into a world that will increasingly confound their parents, and henceforth challenge them to the utmost is pretty stark stuff; that none of us gets out of this one alive is starker still.

Quite apart from one's initial subjective and self-interested responses, gloom-'n'-doom prophecies raise a tough question. What shall we bring in the way of spirit and psyche to times like these? Once at a gathering that consisted of predominantly middle-class white intelligentsia, I listened as Chief Oren Lyons, Faithkeeper of the Onandaga Council, Iroquois Confederacy, described concisely his peoples' lifeway and out of that offered his own counsel for coming times. The wisdom that stuck was that "much will be demanded of those to whom much has been given." (Another thing he said was, "Don't look for mercy where there is none.")

Ours is a pivotal generation. We have to do more now than just live our little lives. We've got to become strong and enduring and respectful and speak to those qualities in one another. Civilization's mishaps have brought us to a time for growing up. Not to despair, and not to hate, and not to frighten, but to engage one another compassionately, to confront lovingly, to embrace the entire reality and not to gainsay the pain are some pieces of the task.

It might be said that spirit is the ultimate flowering of the survival instinct, its sweetest, wisest expression. The movement songs say you can't kill the spirit. Spirit, if it is, is invincible. But people are fragile, flesh and blood, discourageable. External circumstances work differently on different people, and there's no telling why. Where one human may rise in creative outrage against an injustice or senseless act of destruction, another may sink under the waves of deprivation or frustration.

161

Despair, understandable enough in times like these, is perhaps the greatest threat to the flourishing of the human endowment of mind and spirit. Simple ignorance, not knowing even a few details of the infinite wonder of life's patterns in the biosphere and hence not knowing what there is to care for, is another deadly threat. Injustice, racism, alienation, indifference, violence, sexism, loneliness, greed, the destruction of traditional cultures, and the rising levels of suffering that result from ecological deterioration—the list of threats to mind and spirit goes on and on.

Poverty can be a soul killer. Poverty today is underlain by ecosystem degradation. Land-based peoples around the world realize this and are mobilizing to protect their home places. A leader emerges, villagers elsewhere join together to defend their forest or to prevent the dam that would flood their valley. The most celebrated instance of such a mobilization was the Chipko Andolan movement. In 1974 women of Reni, in northern India, interposed their bodies between their forest and the lumberjacks come to fell it. Their protest saved twelve thousand square kilometers of sensitive watershed. When the tree huggers or dam fighters win, and the word of their victory spreads, something budges. A lot of the impetus for such organizing comes from women. This dramatic emergence of women mobilized in defense of their local corner of biosphere is part of the planet-wide challenge to sexism, the pervasive and arbitrary hierarchy of man over woman. However valuable the masculine principle, the world is now desperately in need of greater balance and equity at all levels of society, from the most intimate to the most powerful—hence the hope contained in women's liberation—more energy, a complementary style, another perspective.

Perhaps because our century has seen the use of arms to resolve conflict develop toward a nuclear holocaust, it has also evoked brilliant avatars of nonviolent direct resistance, among whom Mahatma Gandhi and Martin Luther King, Jr., are most known and widely inspiring: great leaders who forswore the use of violence in their struggles for deep social change.

These leaders developed the use of what Gandhi called *satyagraha*—the force of truth itself, which force is an abiding, inclusive love. The unarmed Greenpeace warriors who buzz in their rubber Zodiac boats between whales and whalers; and the Earth

First!ers encamped high in Douglas firs or redwoods in ancient forests to avert logging and clear-cutting are heirs to this tradition of nonviolent direct action in the service of life.

In contrast to the warrior-heroism displayed in such actions stands the patience and anonymity of the Greenham Common women's encampment in England, where since 1981 women have maintained a vigil in opposition to the installation of Cruise missiles. Withstanding continual harassment and the vagaries of weather, these women are not storming the ramparts but solemnly confronting the threat to all life posed by nuclear weaponry. Nonviolent direct action, silent witness, and civil disobedience are immensely powerful and righteous ways to work for change.

*　　*　　*

Deploring humanity's crimes against the planet, and against itself, is not a rhetoric in which we want to get stuck. However, I believe that it is essential for us to acknowledge that we have been capable of atrocities, and to grapple with that capability too as a part of our common, and individual, humanity. Each of us needs to admit and own that reality. Psychologists call this integration.

We need to go forward, out from our experience as the species that is simplifying the biosphere, and begin to live in ways that are sustainable: materially spare, but spiritually rich. And we must take with us humility learned through the failure of our constant, disastrous attempt to dominate nature. Management and control are the pathology of the modern psyche. We could easily be seduced by them yet again as we confront the task of restoring health to a damaged planet.

It seems a little infantile to escape into technofantasies of colonizing Mars with life, or looking to biotechnology to invent "life"-forms that can tolerate our worst pollutions, or relying on some sudden, collective, mass-mediated awakening (which seldom seems to specify any requirement to rein in our wants) to propel us into an abundant millennium.

We must admit that we cannot run the natural world according to the dictates of the ego. This delusion has cost us a half million of our companion species on the planet, and thousands of traditional life-ways. I think that a certain, bounded experience of remorse may be

163

for us seemly and useful. It is a significant part of the reality we must embrace to mature as a species, the inseparable shadow of the joy that life itself engenders.

Penitence might engender in us a passion to heal. Studying the ways of ecosystems, which provides the basis for a faithful restoration of natural landscapes, could become a widespread pastime. It would be a version of inviting the indwelling tutelary spirits of place to tell us how they produce harmony and stability.

The Earth and ecology are not just an idea. The world is real, and human ecological sustainability consists in hundreds of individual physical practices and disciplines, every one of which can be performed as a ritual gesture of respect for the planet that gives us life. Going on litter patrol in the park, starting a compost bucket, and maybe even a community garden to spade it into, learning the name and ethos of that little brown bird in the backyard, traveling on foot through a forest, patching the elbows on a sweater and making do with it a while longer—these are the kinds of disciplines and pleasures that invest a more frugal lifeway with meaning in relation to the Earth.

They are also a means of healing the human spirit—soul-polishing labors and meditations. Engaging in such practice, we can reclaim our intimacy with the Earth, and our community with one another. Imagine the sweetness and freedom in this reunion, the goodness in having a vocation that's a worthy basis for culture.

Learning to provide for ourselves, to care for the places in which we live, and to restore them to biological health; learning to thrive using the renewable energy flows of the elements; studying the natural histories of our home places and discerning in them the outlines of our future self-reliance; consecrating large land areas across every continent as wilderness—shrines where entire ecosystems may continue in their evolutionary destiny—all of these actions aim at becoming native again to place; bioregionalists call it reinhabitation. Reinhabitory humans make love to their home places, bringing forth a wealth of cultures, songs, images, teachings, inventions, musics, and cuisines. Part of the hope in spirit, paradoxically, is coming back to our senses, and being able once again to revel in them.

Remember this: To be alive can feel like such a piece of good fortune as to leave one breathless. To have a pounding heart and nerves that crackle with information and a skeleton strung with muscles strong enough to pump the pedals on the bike or swing the maul to split a little wood; to have eyes to scan the valley and register the subtleness of the color change as September pushes fall forward is the ultimate gift, a consummate luxury. To experience this bliss in life can and should be every human's birthright.

In the invention and regeneration displayed in the wilderness and the renewal displayed in the turning of the seasons, Earth spells her hope for us. In the persistent human yearning after the good, the pure, and the beautiful, and the mounting struggles around the world for a more decent, life-affirming society, are spelled our hopes for ourselves, hopes that are awakened today even in the very young.

*   *   *

One early afternoon, not so very long ago, a cadre of four little girls wound their way up my long driveway to pay a call. Having learned that the home planet is endangered, they had taken it upon themselves to visit all the neighbors in a home-grown canvass to save the Earth. So they had made a creative, positive, and quite touching response to the threat of planet doom. They were giving hand-colored drawings of the Earth, captioned with childishly wise slogans, to each of us privileged to receive a visit. I got one that shows our blue and green planet rising in a field of eight yellow stars. "Save the Earth" reads the headline. "Pitch In. Help Out. A very little goes a long way. If you're interested call. Don't stop trying."

JOHN MUIR

John Muir drew the map which became the basis of the Yosemite National Park bill in 1890, helped found the Sierra Club two years later, personally raised President Theodore Roosevelt's consciousness on matters of forest preservation, and inspired the important, much-used Lacey Antiquities Act of 1906, among many other practical successes. His quantifiable public contributions, indeed, may outstrip those of any other American nature writer.

His philosophical legacy has also been important to American environmentalism, for Muir's writings offer some of the clearest, most experientially vivid accounts we have of the felt unity of man and nature. His descriptions emphasize particulars, often seen scientifically, within an essentially religious stance toward nature and furthermore portray the self of the writer in active continuity with the environment so that what emerges, finally, is a detailed, sensuous, encompassing vision. In his best work, this ecological world view is contagious and inspiring.

Muir was empowered as a philosopher of the wild by spending the summer of 1869, and much of the following five years, in the Sierra Nevada, where he not only experienced the mountain environment in an intense way, but also made a careful, empirical study of its geologic past. During June, July, and August of 1869, at the age of thirty-one, Muir walked the Range of Light, as he came to call it, in the wild area north and east of Yosemite Valley, nominally in the employ of a sheep rancher but actually quite free to explore, to spend days in a meditative state looking out over the dramatic landscape from high points, and above all to absorb a sense of the wilderness's radiant health. After several weeks in the forests and meadows and clean granite uplands, Muir wrote in his journal that the "blessed mountains" were "so compactly filled with God's beauty" that there was no room for *personal* experience, which he now called merely "petty." Clearly, his sense of self had transcended the usual isolative egoism and anthropocentrism characteristic of his culture and ours; he had entered into an ecological identity in which, as he later wrote, "the sun shines not on us but in us. The rivers flow not past, but through us. . . ." Following this "baptismal" summer (Muir's term), he began a serious, extended study of the Sierra's geological history, during which he disproved the then-authoritative cataclysmic theory of the range's topography, showing that the area had in fact been shaped by glaciers. With a series of seven articles outlining his findings, published in the *Overland Monthly* in 1874–1875 (reprinted as *Studies in the Sierra* in 1960), he began to be known as a writer of authority on mountain topics. In newspaper and magazine articles over the next four decades, he combined natural history, personal narrative, poetic-religious reverence for the Earth, and what he called the "conservation gospel" into a characteristic and effective style. During the Progressive Era, when managerial efficiency came to dominate American environmental philosophy (as indeed it still does), Muir articulated a much more profound, ecological outlook.

His most important books, all still in print, are *John of the Mountains: The Unpublished Journals of John Muir* (1938, University of Wisconsin Press); *The Mountains of California* (1894, Sierra Club Books); *My First Summer in the Sierra* (1911, Sierra Club Books); and *Travels in Alaska* (1917, Sierra Club Books). In addition, good writing by Muir has been collected in *Steep Trails* (1918, Houghton Mifflin) and in *John Muir: To Yosemite and Beyond* (1980, University of Wisconsin Press), edited by Robert Engberg and Donald Wesling.

—Thomas J. Lyon

"During the whole two days of storm no idle, unconscious water appeared, and the clouds, and winds, and rocks were inspired with

167

corresponding activity and life. Clouds rose hastily, upon some errand, to the very summit of the walls, with a single effort, and as suddenly returned; or, sweeping horizontally, near the ground, draggled long-bent streamers through the pine-tops; while others traveled up and down Indian Cañon, and overtopped the highest brows, then suddenly dropped and condensed, or, thinning to gauze, veiled half the valley, leaving here and there a summit looming along. These clouds, and the crooked cascades, raised the valley-rocks to double their usual height, for the eye, mounting from cloud to cloud, and from angle to angle upon the cascades, obtained a truer measure of their sublime stature."

—John Muir: To Yosemite and
Beyond

"The air is distinctly fragrant with balsam and resin and mint,—every breath of it a gift we may well thank God for. Who could ever guess that so rough a wilderness should yet be so fine, so full of good things. One seems to be in a majestic domed pavilion in which a grand play is being acted with scenery and music and incense,—all the furniture and action so interesting we are in no danger of being called on to endure one dull moment. God himself seems to be always doing his best here, working like a man in a glow of enthusiasm.

"*June 21.* Sauntered along the river-bank to my lily gardens. The perfection of beauty in these lilies of the wilderness is a never-ending source of admiration and wonder. Their rhizomes are set in black mould accumulated in hollows of the metamorphic slates beside the pools, where they are well watered without being subjected to flood action. Every leaf in the level whorls around the tall polished stalks is as finely finished as the petals, and the light and heat required are measured for them and tempered in passing through the branches of over-leaning trees. However strong the winds from the noon rain-storms, they are securely sheltered. Beautiful hypnum carpets bordered with ferns are spread beneath them, violets too, and a few daisies. Everything around them sweet and fresh like themselves."

—My First Summer in the Sierra

NATURE AND NATURE WRITING:
THREE ANTHOLOGIES

In the beginning, the central, shaping relationship—second only to the bond between mother and child—was that between person and place. Human senses were defined and fine-tuned through interaction with an enveloping natural world—in relationship with the song of birds, tart berries, swirling sandstorms, fleeing animals, and pungent wet earth. The human body is that of a hunter, with muscles and eyesight forged by the vast savannas of our origins. Mind and spirit, too, evolved in interaction with the natural world, as myth found its metaphor in wild nature—in the kindred body and soul of the bear, in the fertility of a seed. Our understanding of how and why the world is—and of ourselves within it—comes, originally, from this seminal relationship between person and place.

"Place" is still where, for the individual, everything connects, where the inner world and outer world come together. The dissonance or harmony of that convergence of "within" and "without" may have a lot to do with our sense of rightness of things. Perhaps it is the imprint in our bodies and minds of an original "rightness"—a remembered intimacy with a wilderness world—that explains, among other things, the continued popularity of nature writing. For this, above all, is what nature writers write about: relationship between person and natural place.

The experience of really knowing, befriending, a place takes time. Months and years of time, along with a measure of solitude and wildness that is hard to come by these days. In an era of mobility and haste—not to mention shrinking wilderness—many people must turn to books, television, and movies to satisfy what seems an innate longing for relationship with the natural world. Though no substitute for personal experience, the best nature writing and film can communicate, vicariously at least, the feelings and insights that are less and less available firsthand.

For all the wonder and beauty of film, nature writing seems to me the truer witness of real relationship. Sixty-minute nature specials conjure a natural world teeming with mating lions and feeding grizzlies. In the real world of nature, it's not like that. In the wild world, animals almost always appear unexpectedly and often only after much waiting. The texture of the wait, the honed attention, the context of wind and changing light and overarching landscape, the synchronistic appearance—the written word, spread over pages, tells the story better and captures subtleties of feeling for which the camera is inadequate.

On this continent, historically, nature writing helped us to begin to know a vast and virgin land—from John Burroughs' natural history essays to John Muir's rhapsodies about the Sierra. Henry David Thoreau and later Aldo Leopold spoke for a respectful and soul-nurturing relationship with the natural world that was a minority view in the days when the frontier attitude of an expansionist nation assumed that forests and wild animals were "commodities" without limit. Today, a new crop of nature writers—many of the best of whom are represented in these anthologies—not only builds on the tradition of Thoreau, Muir, and Leopold, but also, I think, brings something new: a scientifically grounded sense of Earth as a living, breathing organism. This image is remarkably akin to the understanding that we find in ancient myths and oral traditions—of a world throbbing with rich and fragile life, with which humans interact and of which they are a part. Maybe we are coming full circle, to a new understanding of what the natural mind has always known—and through that understanding perhaps we will be able to see our way to the future.

There is no accepted definition of nature writing. These anthologies focus on the nonfiction essay, which usually (though not always) means a person writing about a particular place. On my own nature bookshelves, I also include some anthropology, some fiction and poetry, along with books by the authors represented here and many others. These anthologies, all published since 1987, offer a rich smorgasbord of the genre.

*   *   *

This Incomperable Lande is a "big" book, valuable both as a reference and as a source of pleasure. Lyon's ninety-page introductory section, which reviews and makes sense of the history of nature writing on this continent from his scholarly and environmentally sensitive perspective, is in and of itself worth the price of the book. A seventy-page annotated bibliography at the end of the book is a wonderful resource for further reading. The selections that comprise the middle two hundred pages start with William Wood (1634) and end with John Hay (1987). Lyon has intentionally limited the anthology to "writings by the European-American 'white man' "; his purpose—revealed through his choices and their chronological presentation—is to show the evolution of attitude toward the new land on the part of these "newcomers." In his preface Lyon notes that over time we see something "convergent to the native outlook" developing—by which he means a growing sense of "psychic at-homeness." Lyon introduces each essay with a short sketch of the author and a comment on the importance of his or her writing.

Originally an issue of *Antaeus, On Nature* is the most eclectic and provocative and the least synthesized of these three anthologies. With the help of Annie Dillard, Gretel Ehrlich, Barry Lopez, Robert Finch, John Hay, and Edward Hoagland, Halpern has made thought-provoking selections. The Native American perspective is well represented here, and there are also a couple of Europeans (John Fowles and Italo Calvino). Some of the abridgments (notably of *The Green Man* by Fowles) miss my favorite parts of the originals, but that is a quibble. The contributions of earlier naturalists and writers are covered by contemporary essays on historic figures (Edward Hoagland on John Muir, for instance). An annotated booklist of "natural history" by the advisory editors is fascinating and broad ranging; it shows why a precise definition of nature writing is next to impossible. Even those who know the literature well are likely to find something new and stimulating here.

Stephen Trimble's thirty-page introduction to *Words from the Land* explores the process of nature writing and is highlighted by interviews with nine of the fifteen writers whose work is represented in this book. The insights from these writers, woven together with reflection and further insight by Trimble, are reason enough to add this book to the shelf of any fledgling writer or any aficionado of nature writing. The interviews emphasize that these people think of themselves foremost as *writers*, not naturalists or even environmentalists. In a sense, they are storytellers, telling the story of the place. These authors are all contemporary writers, and, as is the case in all these anthologies, mostly male. But Trimble is the only editor to mention that fact, and to speculate about why there aren't as many (published) women nature writers. (He thinks it may be because this kind of nonfiction essay writing requires solitude, and women may be nervous about being alone in the wilds.) The selections in this book represent well the high literary quality of contemporary nature writing. Many of the essays will be familiar to those who know the genre well, but the interviews, biographical sketches, and introductions add significant value to even the well-stocked library. This is a good first book for those who want to sample contemporary nature writing.

—BARBARA DEAN

This Incomperable Lande: A Book of American Nature Writing
EDITED AND WITH A HISTORY BY THOMAS J. LYON
1989, Houghton Mifflin Company

On Nature: Nature, Landscape, and Natural History
EDITED BY DANIEL HALPERN
1987, North Point Press

Words from the Land: Encounters with Natural History Writing
EDITED BY STEPHEN TRIMBLE
1988, Gibbs Smith

"Most of these materials emphasize in one way or another what seems to me the crucial point about nature writing, the awakening of perception to an ecological way of seeing. 'Ecological' here is meant to characterize the capacity to notice pattern in nature, and community, and to recognize that the patterns observed ultimately radiate outward to include the human observer. This latter dimension may be the key, for it enables an ethical response."

—*Preface*
by THOMAS J. LYON
in *This Incomperable Lande*

"For millennia beyond computation, the sea's waves have battered the coastlines of the world with erosive effect, here cutting back a cliff, there stripping away tons of sand from a beach, and yet again, in a reversal of their destructiveness, building up a bar or a small island. Unlike the slow geologic changes that bring about the flooding of half a continent, the work of the waves is attuned to the brief span of human life, and so the sculpturing of the continent's edge is something each of us can see for ourselves."

—*Wind and Water*
by RACHEL CARSON
in *This Incomperable Lande*

"The bare vastness of the Hopi landscape emphasizes the visual impact of every plant, every rock, every arroyo. Nothing is overlooked or taken for granted. Each ant, each lizard, each lark is imbued with great value simply because the creature is there, simply because the creature is alive in a place where any life at all is precious. Stand on the mesa edge at Walpai and look west over the bare distances toward the pale blue outlines of the San Francisco peaks where the ka'tsina spirits reside. So little lies between you and the sky. So little lies between you and the earth. One look and you know that simply to survive is a great triumph, that every possible resource is needed, every possible ally—even the most humble insect or reptile. You realize you will be speaking with all of them if you intend to last out the year. Thus it is that the Hopi elders are grateful to the landscape for aiding them in their quest as spiritual people."

—*Landscape, History and the Pueblo Imagination*
by LESLIE MARMON SILKO
in *On Nature*

173

"These writers make journeys into the landscape; they enter 'the naturalist's trance.' They weight their journals against their research, spend long hours in libraries, and talk to the experts. They then place themselves in a second trance and translate all this into lovely prose. Within and between the lines lies a wealth of natural history, basics and esoterica, that the reader absorbs painlessly— instructed, invisibly—carried on the buoyant leading edge of the writer's curiosity, skill and enthusiasm."

—*The Naturalist's Trance*
by STEPHEN TRIMBLE
in *Words from the Land*

"A Lakota woman named Elaine Jahner once wrote that what lies at the heart of the religion of hunting peoples is the notion that a spiritual landscape exists within the physical landscape. To put it another way, occasionally one sees something fleeting in the land, a moment when line, color, and movement intensify and something sacred is revealed, leading one to believe that there is another realm of reality corresponding to the physical one but different.

"In the face of a rational, scientific approach to the land, which is more widely sanctioned, esoteric insights and speculations are frequently overshadowed, and what is lost is profound. The land is like poetry: it is inexplicably coherent, it is transcendent in its meaning, and it has the power to elevate a consideration of human life."

—*The Country of the Mind*
by BARRY LOPEZ
in *Words from the Land*

174

News of the Universe: Poems of Twofold Consciousness
CHOSEN AND INTRODUCED BY ROBERT BLY
1980, Sierra Club Books

Poetry provides a form of spiritual nourishment that has been denied to many of us in this era of fast-food, mass media culture. In *News of the Universe* Robert Bly provides us with a healthy sampling of the delights poetry offers the general reader. Like many deep ecologists and feminist theorists, Bly seeks to trace the roots of the dichotomy in Western culture between perception and the objects of perception, saying, "This book asks one question over and over: how much consciousness is the poet willing to grant trees or hills or living creatures not a part of his own species?"

He traces the subject/object dichotomy through time, opening with "Part One: The Old Position," which samples and critiques the rationalist early capitalist underpinnings of seventeenth through nineteenth century European poems. The concluding chapter, "Leaving the House," offers poems that show the richer connections to the natural world provided by older cultures that have not cut away their roots. In between Bly shares eighteenth and nineteenth century poems that attack "the old position," twentieth century poems of twofold consciousness from before and after World War II, and poems that allow the object of the poem to assume its own subjectivity, such as Elizabeth Bishop's poem "The Fish," where upon truly encountering a fish as another creature, the speaker of the poem let it go.

Some theorists speak of a time of second naïveté in our lives, during which we are able both to accept the humanly given nature of ritual and to accept ritual as a way of ordering our lives. Poems of twofold consciousness help us sustain a vision of our true relationship to the planet.

—CAROL DORF

"What I've called the Old Position puts human reason, and so human beings, in the superior position. The Old Position may be summed up, or oversimplified, this way: Consciousness is human, and involves reason. A serious gap exists between us and the rest of nature. Nature is to be watched, pitied, and taken care of if it behaves."

"I think that the Westerner, after centuries of planning— successful planning—loses touch with nature so much that he loses touch with her dark side also. . . . Literature and art attempt to reopen channels between human beings and nature, and to make our fear of her dark side conscious, help us to see her without fear, hatred, or distance. . . ."

"To feel the contrast between our contemporary experience when we look at an object or a hillside, and the experience that is possible when an 'opened' human being does that, we have to go far back into the past of the human race."

"I don't want to imply too great a unity in nature's consciousness. The varieties of consciousness inside things can be suggested by the way a ray of light unfolds into a certain abundance as it passes through a prism, from red, through yellow and green and blue to violet. One could say that the unconsciousness inside the badger belongs to the violet band, with much night-vision and melancholy. Trees belong elsewhere in the spectrum."

The Dream of the Earth
THOMAS BERRY
1988, Sierra Club Books

In his compelling book *The Dream of the Earth*, Thomas Berry, a prolific writer on environmental ethics, has taken a fascinating plunge into history and thoughtfully traced man's increasing alienation from nature.

In chapter after chapter, Berry explores the evolving way of life of our human species—our burgeoning technologies, our economic concepts, our notions about education and our culture, relating them all to an idea of progress that has fueled . . . a commitment to unlimited growth . . . "achieved by plundering processes."

Through the explosion of scientific knowledge and intensive industrialization, it is clear that we are changing the planet's natural order. We are altering the chemistry of the planet, the integrity of the great hydrological cycles and the ozone layer that shields us from harmful rays, and we are inundating our air, water, and soil with toxic substances.

"The essence of the human dilemma," Berry writes, is that "the energy evoked by the ecological vision has not been sufficient to offset the energies evoked by the industrial vision—even when its desolation becomes so obvious as it is at the present time."

But *The Dream of the Earth* is not just a tale of woe. In the growing scientific evidence of the havoc human activity has wreaked on the natural resources that have sustained human life and made it rich and varied, Berry sees the beginnings of a new "vision." He finds a ray of hope in the budding environmental movement, a new listening as the Earth's systems speak to us "through the deepest elements of our nature, our genetic coding." Berry warns, though, that "the order of magnitude of the changes that are needed is great. We are not concerned with some minor adaptation but with the most serious transformation of human-earth relations that has taken place at least since classical civilizations were founded." The goal that must be attained is the creation of a new cultural context in which people may shed their anthropomorphic view of life and replace it by a biocentric one, and learn to live as a harmonious part of a continually self-renewing natural world.

—JOANNA D. UNDERWOOD

"Our difficulty is that we are just emerging from a technological entrancement. During this period the human mind has been placed within the narrowest confines it has experienced since consciousness emerged from its Paleolithic phase. Even the most primitive tribes have a larger vision of the universe, of our place and functioning within it, a vision that extends to celestial regions of space and to

177

interior depths of the human in a manner far exceeding the parameters of our own world of technological confinement."

"In every country, then, a mystique of the land is needed to counter the industrial mystique. This mystique must be associated with the three basic commitments of our times: commitment to the earth as irreversible process, to the ecological age as the only viable form of the millennial ideal, and to a sense of progress that includes the natural as well as the human world."

The Chalice and the Blade: Our History, Our Future
RIANE EISLER
1987, Harper & Row

In *The Chalice and the Blade*, Riane Eisler tells a story that we badly need to understand. Humanity has not always been organized in the male-dominated, hierarchical society that exploits and destroys its natural environment. A mere five to six thousand years ago, the female-inspired epoch that we know as the Neolithic period was climaxing in all the centers of human culture that later became patriarchal civilizations. Eisler has focused attention on the creative and peaceful character of the Neolithic matrifocal societies contrasted with the succeeding violent civilizations. Her point is that if there once was this other way of being human, does it not seem possible that we can evolve out of patriarchy into a nonhierarchical, ecological society with a coordinate relationship between the sexes?

Eisler's success in capturing the imagination comes from her narrative choice of a particular version of this ancient worldwide change. She describes the melodrama of invasion of peaceful, matrifocal Old Europe settlements by destroying patriarchal, Indo-European warriors. Unfortunately, this version of this change applies only to Greece and India and not to the other seven first civilizations. In those other places, there were no clear patriarchal villains, and the change came from within the matrifocal tribal culture itself, raising more difficult anthropological-sociological questions than Eisler tackles.

Now that her successful publicism has riveted the attention of a wider-than-feminist audience on what would otherwise be a hopelessly arcane question, we are freed to explore the exact psychological and institutional change that would take us beyond patriarchy, an exploration that pioneer Eisler only begins to suggest in this, the first volume of a contemplated trilogy.

—PHIL HOLLIDAY

"As far east as Harappa and Mohenjo-Daro in India, large numbers of terra-cotta female figurines had earlier been found. These, too, as Sir John Marshall wrote, probably represented a Goddess 'with attributes very similar to those of the great Mother Goddess, the Lady of Heaven.' Goddess figurines have also been found in European sites as far west as those of the so-called megalithic cultures who built the huge, carefully engineered stone monuments at Stonehenge and Avebury in England. And some of these megalithic cultures went as far south as the Mediterranean island of Malta, where a giant ossuary of seven thousand burial sites was apparently also an important sanctuary for oracular and initiary rites in which, as James writes, 'the Mother-goddess probably played an important part.'

"Gradually a new picture of the origins and development of both civilization and religion is emerging. The Neolithic agrarian economy was the basis for the development of civilization leading over thousands of years into our own time. And almost universally, those places where the first great breakthroughs in material and social technology were made had one feature in common: the worship of the Goddess."

"In sum, in contrast to the still prevailing view of power as the power symbolized by the Blade—the power to take away or to dominate—a very different view of power seems to have been the norm in these Neolithic Goddess-worshiping societies. This view of power as the 'feminine' power to nurture and give was undoubtedly not always adhered to, for these were societies of real flesh-and-blood people, not make-believe utopias. But it was still the normative ideal, the model to be emulated by both women and men.

"The view of power symbolized by the Chalice—for which I propose the term *actualization power* as distinguished from *domination power*—obviously reflects a very different type of social organization from the one we are accustomed to. We may conclude from the evidence of the past examined so far that it cannot be called matriarchal. As it cannot be called patriarchal either, it does not fit into the conventional dominator paradigm of social organization."

"The move to a new world of psychological and social rebirth will entail changes we cannot yet predict, or even envision. Indeed, because of so many failures following earlier hopes for social betterment, projections of a positive future elicit skepticism. Yet we know that changes in structure are also changes in function. Just as one cannot sit in the corner of a round room, as we shift from a dominator to a partnership society, our old ways of thinking, feeling, and acting will gradually be transformed.

"For millennia of recorded history, the human spirit has been imprisoned by the fetters of androcracy. Our minds have been stunted, and our hearts have been numbed. And yet our striving for truth, beauty, and justice has never been extinguished. As we break out of these fetters, as our minds, hearts, and hands are freed, so also will be our creative imagination."

Woman and Nature: The Roaring Inside Her
Susan Griffin
1978, Harper & Row

Woman and Nature: The Roaring Inside Her is one of those books that cannot be read in a sitting. Because it makes you weep. Many passages overwhelmed me as Susan's words put shape to my deepest understandings, so much so that I often had to put the book down, lift my head up, and see and feel the world anew, through the mist of tears. This experience, time and again, evoked the roaring inside me, that is inside us all, that must be heard to heal the rift between spirit and matter, humanity and nature, woman and man. Susan weaves feeling and thought deliberately to infuse the text with the very stuff of humanity's separation from nature. There are two voices, set in a kind of dialogue—one the authoritative, objective, paternal voice; the other, set in italics, that of woman and nature, passionate and embodied.

The book is designed to stir this roaring as it traces Western civilization's history, showing how woman and nature have been regarded by patriarchy—as existing for the use of and abuse by the self-interested. Page after page poignantly reveals the callousness with which "matter"—in the patriarchal perspective, all "other"—has been manipulated and controlled in order to quell man's fear of his own mortality. How uniform is his attitude toward women and cows, trees (read lumber) and efficient secretaries, show horses and the ideal female human. We are told how we have become utterly dependent, how we have been denied our bodies, how rationality has seemingly ruled the universe. Yet, this universe shudders; the Earth grows impatient; there is an inevitable remembering.

The reader is taken to the "age of her resonance," toward "the separate rejoined," where the dreams women have always dreamed of how things could be must now be spoken to remember that, like our mothers before us, we are made from this Earth and we can no longer bow to tyranny. The women are speaking now for all that has been unspoken. This book helps us find the words to name the unnamed; it gives us the temerity to roar for the love of this Earth. Please read it.

—JUDITH PLANT

"We say he should have known his action would have conse-quences . . . that one act cannot be separated from another. And had he seen more clearly, he might have predicted his own death. How if the trees grew on the hillside there would be no flood. And you cannot divert this river. We say look how the water flows from this place and returns as rainfall, everything returns, we say, and one

thing follows another, there are limits, we say, on what can be done and everything moves. We are all a part of this motion, we say, and the way of the river is sacred, and this grove of trees is sacred, and we ourselves, we tell you, are sacred."

"We shall tell you who gained and who lost In this way *for there were those* the fathers *who held* knew that their names *and those who did not* would live on *those who were known* and that the great estates *and those* testifying to their glory and fame *who were unknown* would live on *those whose labor were vanished* and that the power which spread from those holdings of land *those whose labor* would continue, generation after generation *like the labor of the fields, of the soil,* to be great in the minds of the living *would pass like the passing of breath from the living."*

"Barely seen, soundlessly surrounding him, with hardly a breath of evidence, all he has burned, all he has mined from the ground, all he cast into the water, all he has torn apart, comes back to him . . . *Who speaks to us in our dreams. Sings in our blood and will not be still there.* Every attempt he makes to order this world decreases his space. The dimensions of his life are filled with ghosts *Making us grieve for no apparent reason. Making us rage for no visible reason.* filled with shadows, with tiny reminiscences . . . *This fury in us that will not die, who has captured our bodies, who claims to have been with us for years. Who is making us see what we have not seen before. He is haunted: all his victims speak in his body. He cannot escape pollution, there is no way for him to be free of these ghosts."*

Whatever Happened to Ecology?
STEPHANIE MILLS
1989, Sierra Club Books

Whatever Happened to Ecology? is a moving personal remembrance and reflection on the history of the environmental movement from

182

the viewpoint of an activist whose work spans the late 1960s to the present. The book begins with Stephanie Mills's famous graduation speech in 1969, when she shocked her elders by renouncing motherhood—a move that swiftly catapulted her into the ecology and population-control movements. The book ends with her life in rural Michigan and the continuing daily, and often unsuccessful, struggle to live in harmony with the Earth.

In between, there are fascinating portraits of environmental activists like David Brower, Joan McIntyre, Stewart Brand, John and Nancy Jack Todd, and many other less well-known figures. Deep ecology, bioregionalism, and spirituality receive subtle and loving treatment.

Mills refuses to accept easy oversimplifications and unflinchingly faces the complex and difficult task of human beings trying to live morally in a world of contradictions. An extraordinary honesty characterizes this beautifully written book.

In one section Mills describes the building of a home in Michigan and how, despite the best intentions in the world, the compost toilet was abandoned for a more regular low-flush toilet. In another section, she shows how even the most determined middle-class environmentalists can wake up one day to realize how much like yuppies they've become without even noticing.

While there are beautiful descriptions of nature here, it is Mills's insights about human beings that seem uniquely perceptive. She writes: "Some part of me, I confess, does not rejoice to see new couples start new families. . . . Down in the depths, there is a part of me that regards humans as Gaia's worst affliction. It is a horrifying soul content, but I must own it. It's the kind of attitude that, writ large, becomes a Nazi. My higher self, the self that for ethical, if not rational, reasons chooses hope, knows better." There, rendered in one startling paragraph, is the entire dispute between social and deep ecology, but as an internal battle inside each person. Mills writes about hope and fear, about the continuing human need for heroics and dreams, as well as the dark side of the tendency toward self-hatred. By the end, Mills puts forward the hope that life's diversity, even if reduced and severely threatened by humans, will survive. Whether this optimism proves prescient or illusory, it is something we need to hear. There is no way for us to be agents of planetary healing if we believe we are fundamentally evil.

—MARGOT ADLER

"We humans seem to have been fated by anatomy and then consciousness to imagine ourselves as being uniquely independent from the ecological ground of being. Knowing what we know now, we are essentially incapable of going back to that undifferentiated awareness. Besides, in nature there's no 'back' to go to. All but the tiniest fraction of the wilderness that birthed us is gone. So we must conceive a way to go forward. I think that this will consist in, among other things, discovering the grace of death, in making peace with it. Heading into the cycle and accepting our destiny also to be prey, and to give life again to other organisms."

"The woods are not exactly full of morels, but diligent searching is sometimes rewarded. Mushroom hunting, for the semirationalist, involves wandering short-sightedly in the woods, scanning the ground at your feet, hoping to discern the matte ebon fungus among its cognate shapes on the forest floor—the tiny lean-tos of leaves pushed up by fiddleheads in their striving to the light, the absorbent dark butts of toppled saplings, the shadows inside fallen beechnuts—all of these living and dying elements of the surface can hook the eye into a moment's attention. Seeing the evident relations and identities of things—their proportion, their color, their contours—discloses harmony, mutual aid, survival of the subtlest."

Taking Charge of Our Lives: Living Responsibly in the World
AMERICAN FRIENDS SERVICE COMMITTEE OF SAN FRANCISCO
EDITED BY JOAN BODNER
1984, Harper & Row

Here's a useful, principled, and challenging book. Incorporating the collective wisdom of scores of people associated with the now-defunct Simple Living Program of the American Friends Service Committee in San Francisco, *Taking Charge of Our Lives* is a second-generation handbook on the whys and hows of simple living. In addition to offering many practical suggestions about what individuals, families, and households can do to arrive at a greater material simplicity, and hence economic freedom, the book challenges the reader to explore his or her investment in the privatism that is so conducive to excessive consumption. Herein are plenty of inspiring instances where ordinary folk took action to withdraw their cooperation from or to confront the system; strategies for meeting basic needs for food, shelter, right livelihood, and child care; and extensive bibliographies and resource directories.

"The more self-reliant life is not a life of convenience," say the authors. Nor is it a life of collusion in the economic injustice and apolitical numbness that accompany the American standard of living. *Taking Charge of Our Lives* doesn't minimize the corporate and state roles in deepening global and local inequity, but it honorably cleaves to the principle of nonviolent direct action as the ultimate means for changing these realities. "Simple in means, rich in ends" is the nutshell version of both deep ecology and voluntary simplicity. The richness lies in the empowerment and flourishing of the self in sustainable communities. The need for this has never been greater.

—STEPHANIE MILLS

"Changing long-established patterns of work and material consumption is a step-by-step process, even for someone like Gandhi. In recalling his decision to discard wealth and possessions in order to serve the people, Gandhi stated: 'I cannot tell you with truth that, when this belief came to me, I discarded everything immediately. I must confess to you that progress at first was slow. And now, as I recall those days of struggle, I remember that it was also painful in the beginning. . . . But as days went by, I saw that I had to throw overboard many other things which I used to consider as mine, and a time came when it became a matter of positive joy to give up those things.'

"You and I, if considering a commitment to simpler material living, may not wish to throw overboard as much as Gandhi did. The world's resources can probably support a higher standard of material comfort than Gandhi adopted for all of the world's people on a sustainable basis. Rather, we can adopt an attitude of stewardship and partnership with the natural world and with our sisters and brothers everywhere, by getting in the habit of asking questions about each choice instead of taking the status quo for granted. Is getting our clothes 'white' more important than having unpolluted waters? Is the opportunity to purchase exotic foods more important than hungry families having control over enough land to grow basic foods? Is the use of aluminum in manufacturing soft drink cans more important than using that same aluminum for manufacture of light bicycle frames for basic transportation in hilly terrain? Each choice by each individual may seem small, but the combined effect creates the material, social, and spiritual climate that we live in."

"Consuming the World—Queries

1. Do American consumption patterns provide a good example for other nations?
2. Do I think the 'American way of life' can be maintained at its present level of resource consumption indefinitely? Who would benefit? Who would suffer?
3. What relationship does militarism and military spending have to global resource use and distribution?
4. Are the principles for sustainable and ecologically balanced development different for 'more developed' societies and 'less developed' societies?

5. What burdens do overall American consumption patterns place on poorer people at home and abroad?
6. Are we individually responsible for injustice that is built into the world economy?
7. How can living more simply affect global economic conditions?

"The patterns of consumption largely taken for granted by the majority of the U.S. population bear a heavy responsibility for the perpetuation of poverty. This occurs directly through our own impact on global resources through consumerism, military spending, and the corporate investment that supplies and profits from these. It occurs indirectly through the spread of a consumer culture based on American and European patterns, as popularized through movies and advertising."

Small Is Beautiful: Economics As If People Mattered
E. F. SCHUMACHER
1973; 1975, Harper & Row

E. F. "Fritz" Schumacher and his most famous book, *Small Is Beautiful*, occupy a key niche in the bibliography of the global environmental movement. Most early environmentalists, myself included, confronted an impenetrable wall of traditional economic theories—from left to right—which held our environmental values—clean air and water, biological diversity, healthy soils, and wilderness—to be essentially worthless.

Then, in 1973, *Small Is Beautiful* burst onto the environmental scene in Britain and the United States. This book gave us all the arguments we needed to face down the legions of economists, and their short-sighted cost-benefit analyses, wherever we encountered *their* arguments for the status quo—at public hearings, on campuses, and in our community-organizing work.

As I reread *Small Is Beautiful,* I was still amazed at its clarity and foresight—identifying virtually all the key issues that have emerged from the environmental movement over the past fifteen years, from nuclear energy to the debate over appropriate uses of technology. Schumacher offered the broader contexts so necessary to understanding these issues as well. By addressing narrow, short-term economic theories and challenging the shibboleths of economic growth at all costs, Schumacher's goal was nothing less than epistemological demolition and reconstruction, or, as we might say today, changing the paradigm.

Schumacher predicted the breakdown of chemical and energy-intensive agriculture in his chapter "The Proper Use of Land," urging soil conservation and organic methods instead. He always walked his talk and founded the British-based Soil Association, as well as the Intermediate Technology Development Group in London.

Schumacher's early discussions of "the economics of permanence" undergird all of today's debates about sustainability, and in addressing development issues for so-called Third World countries, Schumacher was a genuine globalist, rather than merely a global ecologist. Today's issues of acid rain, atmospheric CO_2, and forest destruction in the North and the South would not surprise him. Not only did he predict them, but he also had some of the answers: "Plant trees!" In fact, in one of his reports in the 1960s to leaders in India, he urged government tree-planting programs be sponsored throughout the nation as the best, most systemic form of development. He is still revered in India today, along with Mahatma Gandhi, one of his mentors.

Indeed, *Small Is Beautiful* should be required reading, not only for the next generation of environmentalists, but also for those theorizing on America's current decline in productivity and international standing, as well as those debating our current widespread cultural confusion over our values, ethics, and goals. It is one of those rare books that improves with age.

—HAZEL HENDERSON

"Simplicity and non-violence are obviously closely related. The optimal pattern of consumption, producing a high degree of human satisfaction by means of a relatively low rate of consumption, allows people to live without great pressure and strain and to fulfil the primary injunction of Buddhist teaching: 'Cease to do evil; try to do good.' As physical resources are everywhere limited, people satisfying their needs by means of a modest use of resources are obviously less likely to be at each other's throats than people depending upon a high rate of use. Equally, people who live in highly self-sufficient local communities are less likely to get involved in large-scale violence than people whose existence depends on world-wide systems of trade."

"Almost every day we hear of mergers and takeovers; Britain enters the European Economic Community to open up larger markets to be served by even larger organisations. In the socialist countries, nationalisation has produced vast combines to rival or surpass anything that has emerged in the capitalist countries. The great majority of economists and business efficiency experts supports this trend towards vastness.

"In contrast, most of the sociologists and psychologists insistently warn us of its inherent dangers—dangers to the integrity of the individual when he feels as nothing more than a small cog in a vast machine and when the human relationships of his daily working life become increasingly dehumanised; dangers also to efficiency and productivity, stemming from ever-growing Parkinsonian bureaucracies.

"Modern literature, at the same time, paints frightening pictures of a brave new world sharply divided between *us* and *them*, torn by mutual suspicion, with a hatred of authority from below and a contempt of people from above. The masses react to their rulers in a spirit of sullen irresponsibility, while the rulers vainly try to keep things moving by precise organisation and coordination, fiscal inducements, incentives, endless exhortations and threats."

"It is doubly chimerical to build peace on economic foundations which, in turn, rest on the systematic cultivation of greed and envy, the very forces which drive men into confict.

"How could we even begin to disarm greed and envy? Perhaps by being much less greedy and envious ourselves; perhaps by resisting the temptation of letting our luxuries become needs; and perhaps by even scrutinising our needs to see if they cannot be simplified and reduced. If we do not have the strength to do any of this, could we perhaps stop applauding the type of economic 'progress' which palpably lacks the basis of permanence and give what modest support we can to those who, unafraid of being denounced as cranks, work for non-violence: as conservationists, ecologists, protectors of wildlife, promoters of organic agriculture, distributists, cottage producers, and so forth? An ounce of practice is generally worth more than a ton of theory."

Only One Earth: The Care and Maintenance of a Small Planet
BARBARA WARD AND RENÉ DUBOS
1972, W. W. Norton and Company

Our Common Future
THE WORLD COMMISSION ON ENVIRONMENT AND DEVELOPMENT
1987, Oxford University Press

Most books seeded by international conferences, born of multiple authors and blessed by large impersonal organizations, turn out gutless and die young. But a few transcend their joyless births and leave a lasting mark on society and the institutions that sponsored them. *Only One Earth* (an unofficial report commissioned by the 1972 U.N. Conference on the Human Environment) is one, and *Our Common Future* (released in 1987 by the U.N. Commission on Environment and Development) looks to be another.

Only One Earth opened a whole generation's eyes to the increasingly global nature of environmental threats, the unbreakable links between human poverty and environmental degradation, and the need to size up the genie of technology fast. Passionate, wise, and off-handedly learned, the authors proved prophetic in their concern about such issues as the greenhouse effect (yes, 1972!), the Los Angelization of the planet, the growing gap between rich and poor, and the fast-approaching end of the fossil fuel era. More important, they wrestled with the big environmental questions: How does modern science confirm the wisdom of the ancients? Where is technology taking us? And what, spiritually and economically, do we have to lose if we stay on the course we chose—if only by default—around mid-century? What other "report" ever ended like this: "Is [Earth] not a precious home for all of us earthlings? Is it not worth our love?"

Our Common Future never scales such heights or plumbs such depths. Yet this measured but still-shocking assessment and new "global agenda for change" can be considered a well-wrought game plan for making good on the promise of "one Earth" before it's too late. More timely than timeless, the book reached best-seller status in Europe, where sensitivity to the Third World's problems is greater than it is in the United States and where the greening of governments is farther along. It deserves far more attention here.

—KATHLEEN COURRIER

"Rich and poor, developed and developing, industrialized and pretechnological—all are enmeshed in myriad webs of trade, communication, and influence, all are struggling to adapt the technological order to truly human ends, all are involved in the welfare and survival of their fellow communities, all must inescapably share a single, vulnerable biosphere."

"No nations, on their own, can offset the risk of deepening disorder. No nations, acting singly or only with their own kind, rich or poor, can stave off the risk of unacceptable paternalism on the one hand or resentful rejection on the other."

"Our prophets have sought it. Our poets have dreamed of it. But it is only in our own day that astronomers, physicists, geologists, chemists, biologists, anthropologists, ethnologists, and archaeologists have all combined in a single witness of advanced science to tell us that, in every alphabet of our being, we do indeed belong to a single system, powered by a single energy, manifesting a fundamental unity under all its variations, depending for its survival on the balance and health of the total system."

—Only One Earth

"Economic growth and development obviously involve changes in the physical ecosystem. Every ecosystem everywhere cannot be preserved intact. A forest may be depleted in one part of a watershed and extended elsewhere, which is not a bad thing if the exploitation has been planned and the effects on soil erosion rates, water regimes, and genetic losses have been taken into account. In general, renewable resources like forests and fish stocks need not be depleted provided the rate of use is within the limits of regeneration and natural growth. But most renewable resources are part of a complex and interlinked ecosystem, and maximum sustainable yield must be defined after taking into account system-wide effects of exploitation."

"*Changing the Quality of Growth*
"Sustainable development involves more than growth. It requires a change in the content of growth, to make it less material- and energy-intensive and more equitable in its impact. These changes are required in all countries as part of a package of measures to maintain the stock of ecological capital, to improve the distribution of income, and to reduce the degree of vulnerability to economic crises."

—Our Common Future

Green Politics: The Global Promise
FRITJOF CAPRA AND CHARLENE SPRETNAK
1984; 1986, Bear and Company

The Green Alternative: Creating an Ecological Future
BRIAN TOKAR
1987, R. and E. Miles

Green Politics has been credited with stimulating the growth of a Green political movement in the United States, which was certainly its intention. (Just goes to show you the importance of an idea.) In their 1984 volume, Capra and Spretnak reported on the German Greens' political program. The German Greens' "four pillars"— ecology, social responsibility, grass roots democracy, and nonviolence—together constitute a powerful new political idea, indeed.

Written to provide environmental activists and interested bystanders with an introduction to a most hopeful European political phenomenon—the emergence of Green parties as an electoral force—*Green Politics* begins with a description of the seating in 1983 of twenty-seven Green parliamentarians in West Germany's Bundestag.

This was a signal event, possibly the beginning of a politics for the real world, a biological world, a world in which violent confrontation wisely is abjured, a world in which the concerns of all people— not just elites—are reckoned. Capra and Spretnak recount this history-in-the-making, the emergence of Green activity in other nations, and are entirely sensible of the hope it represents.

The possible significance of Green politics on American soil is a subject of *The Green Alternative*, Brian Tokar's fine holistic manifesto. Rooted in historical understanding, attentive to current happenings, and committed to a creative vision for the future, Brian Tokar's eloquent, lucid, idealistic little volume is suffused with the conviction that an ecologically sound society can and must develop

193

here. Tokar pulls no punches in assessing the failings and dangers of present governance and culture. He also provides examples of greening taking place in the here and now, such as the conversion, by the city of Arcata, California, of an old shoreline dump into a wildlife sanctuary. Clear and intelligent, gentle and decided, *The Green Alternative* is peerless.

—STEPHANIE MILLS

"Participatory democracy is not for the impatient. The development of a party program, which also stands as a campaign platform, involves extensive consultation with the grassroots membership. The Green party in the state of Hesse, for instance, selected a program committee who invited suggestions from all members on the issues of peace; employment and economy; energy; the environment; city planning, living places, traffic management; democratic rights; culture and education; women; children and young people; elderly people; discrimination against minority groups; and health. They also included an appendix presenting four global alternatives for the future. When the committee members had compiled the ideas from the general membership, they sent the program back to the one hundred local groups in their state, asking for refinements, changes, and further ideas. After the committee had incorporated the changes, a statewide party assembly met for six consecutive weekends to discuss the various points in contention and to agree upon a final draft of the program."

—Green Politics

"One Green group in the central hills of Vermont has proposed the development of alternative economic experiments in tandem with the creation of a Community Congress of existing community institutions. By merging these two efforts, they hope to encourage a new strength of working relationships among the different sectors of both the alternative and traditional communities. In Boston, some Greens are working with activists in the black community to turn an abandoned state hospital site into a center for education and experiments in urban agriculture and alternative technologies. The center would employ local people, teach new skills necessary for self-reliance and provide healthy food for inner-city residents. At the same time, the project would help curtail land speculation and commercial over-development in the midst of a residential neighborhood by averting plans to turn the site into a new high-technology industrial park."

—The Green Alternative

Rules for Radicals
Saul Alinsky
1971; 1989, Random House

"The one thing certain is that masses of middle-class Americans are ready to move toward confrontation with Corporation America," so Saul Alinsky, the father of community organizing, wrote in 1971. Let's hope so.

Two decades later, as young organizers prepare themselves for battle, the Alinsky primer remains the bible for organizers bent on greater social and political justice.

A quick two-hour read provides the combined wisdom of Machiavelli, Lord Acton, Edmund Burke, Freud, Socrates, Moses, John L. Lewis, Clarence Darrow, and just about anyone else who ever thought about the powerful and the powerless. All this and Alinsky's own commonsense original genius.

Rules for Radicals describes poor, working, and middle-class organizing campaigns and even touches on the early development of corporate and proxy campaigns being fought today.

Irreverent, serious, historical, inspirational, and totally contemporary, *Rules for Radicals* should still be number one on the reading list to educate the young and refuel the veterans.

—Barbara Shailor

"The internecine struggle among the Haves for the individual self-interest is as shortsighted as internecine struggle among the Have-Nots. I have on occasion remarked that I feel confident that I could persuade a millionaire on a Friday to subsidize a revolution for Saturday out of which he would make a huge profit on Sunday even though he was certain to be executed on Monday."

"The answer I gave the young radicals seemed to me the only realistic one: 'Do one of three things. One, go find a wailing wall and feel sorry for yourselves. Two, go psycho and start bombing—but this will only swing people to the right. Three, learn a lesson. Go home, organize, build power and at the next convention, *you be the delegates.*'

"Remember: once you organize people around something as commonly agreed upon as pollution, then an organized people is on the move. From there it's a short and natural step to political pollution, to Pentagon pollution.

"It is not enough just to elect your candidates. You must keep the pressure on. Radicals should keep in mind Franklin D. Roosevelt's response to a reform delegation, 'Okay, you've convinced me. Now go on out and bring pressure on me!' Action comes from keeping the heat on. No politician can sit on a hot issue if you make it hot enough."

Ecotopia
ERNEST CALLENBACH
1975, Banyan Tree Books*

Ecotopia Emerging
ERNEST CALLENBACH
1981, Banyan Tree Books*

Casual sex, organic gardens, and all-terrain bikes. Fast trains, solar heat, and no nukes. Recycling, tree hugging, and clean rivers. . . . Where on Earth is that, Sweden?

* Banyan Tree Books, distributed by Bookpeople, 2929 Fifth Street, Berkeley, California 95710

No; Ecotopia . . . California, Oregon, and Washington. No longer part of the United States . . . an earthly Eden populated by a suddenly wise and car-less people.

*   *   *

It was a different world, back in 1975 when *Ecotopia* was published. Jimmy Carter was unknown, as were AIDS, Ayatollah, and ozone. And the Iron Curtain was still solidly in place. Sixties idealism was still alive, the selfishness of Reaganism unborn.

For those reasons, Ernest Callenbach's environmental classic seems a bit dated—when read today. But on reflection, the man's vision can't be seen as anything but impressive. He took an ecological want list and turned it into a story, a tale in which Americans (hear that, World? *Americans!*) had somehow changed their wasteful ways and become Earth stewards.

Here they are, the people who stole a continent, paved its buffaloes, leveled Hiroshima, and wrecked Vietnam, suddenly turned Green! Callenbach wanted to paint a picture of total environmental commitment, something perhaps akin to the spirits of Earth First! and deep ecology, as opposed to the self-serving shallowness of corporate "environmentalists." He even took on the problem of our genetic war lust and dealt with that.

Science fiction? In a way. In a down-home way.

Ecotopia's sister book, *Ecotopia Emerging*, published six years after the best-seller appeared, goes back and tells how Ecotopia evolved as a nation, how it seceded from the United States. In many ways a more sophisticated novel, it makes the first book more believable. It gives it roots, and that can't help but give it new life just when we need so badly to think about a gentle alternative to what we're doing.

—MALCOLM WELLS

"By the time you notice you are underway in an Ecotopian train, you feel virtually no movement at all. Since it operates by magnetic suspension and propulsion, there is no rumble of wheels or whine or vibration. People talk, there is the clink of glasses and teacups, some passengers wave to friends on the platform. In a moment the train seems literally to be flying along the ground, though it is actually a few inches above a trough-shaped guideway. . . . You get a fair view of the countryside at this speed, which translates as about 225 miles per hour. And we only attained that speed after about 20 minutes of crawling up and over the formidable eastern slope of the Sierra Nevadas, at what seemed less than 90 miles per hour!"

—Ecotopia

"TOP SECRET
TO: THE PRESIDENT
COPIES TO ALL STATION CHIEFS
An underground group, probably affiliated with the 'Ecotopian' movement, whose overt arm is the Survivalist Party, may have obtained supplies of weapons-grade uranium sufficient to build several nuclear devices. Agents believe the group's purpose is political blackmail of the federal government in order to facilitate secession of the Ecotopia region—Washington, Oregon, and California as far south as the Tehachapi range. See previous memoranda No. A3564 and No. A3992."

—Ecotopia Emerging

Gandhi on Nonviolence
EDITED BY THOMAS MERTON
1965, New Directions

The Power of the People: Active Nonviolence in the United States
EDITED BY ROBERT COONEY AND HELEN MICHALOWSKI
1987, New Society

I read Thomas Merton's *Gandhi on Nonviolence* before I began The Veteran's Fast for Life in 1986. Merton's introduction is the best analysis of Gandhi's views available in twenty pages. The rest of the book is a selection of salient quotes from Gandhi's great work, *Nonviolence in Peace and War*, on various themes such as spirituality, politics, and purity. This slim little volume illuminates so much of the essential in Gandhi: his synthesis of Eastern and Western ideas, including Christianity, with Eastern religions; his insistence that religion and politics are inseparable; the core wisdom that nonviolence comes naturally from inner unity and peace. I read it again and again and find it helps me penetrate deeper into the profound nature of nonviolence each time.

The Power of the People is the best history book of nonviolent dissent in the United States I know of. It should be required reading in every high school in the country. Had I read this book when I was growing up, I would have had a completely different view of U.S. history. Through wonderful photographs and well-written text, it very concretely describes nonviolent actions, people, and events dating back more than two hundred years to the early pacifists of the colonial period, before the word *nonviolence* was even used. The many images—of suffragettes, conscientious objectors, Greenpeace actions, antinuclear protesters, civil rights marches—convey with deep conviction the fact that violence is not the only tradition we inherit.

Gandhi taught that love is the law of our being. Nonviolence is just another word for love, which is ultimate respect for all life. When we get away from understanding that we are all interconnected in the fabric of life, that we can't spit on the Earth without spitting on ourselves, we make choices that disregard life and we ultimately pay the price.

—BRIAN WILSON

"*Ahimsa* (non-violence) is for Gandhi the basic law of our being. That is why it can be used as the most effective principle for social action, since it is in deep accord with the truth of man's nature and corresponds to his innate desire for peace, justice, order, freedom, and personal dignity. Since *himsa* (violence) degrades and corrupts man, to meet force with force and hatred with hatred only increases man's progressive degeneration. Non-violence, on the contrary, heals and restores man's nature, while giving him a means to restore social order and justice. *Ahimsa* is not a policy for the seizure of power. It is a way of transforming relationships so as to bring about a peaceful transfer of power, effected freely and without compulsion by all concerned, because all have come to recognize it as right.

"Since *ahimsa* is in man's nature itself, it can be learned by all, though Gandhi is careful to state that he does not expect everyone to practice it perfectly. However, all men should be willing to engage in the risk and wager of *ahimsa* because violent policies have not only proved bankrupt but threaten man with extinction."

—*Gandhi on Nonviolence*

"Although there were pacifist organizations prior to the Civil War, pacifism was still conceived as a personal creed and direct action was carried out on a more or less individual basis. Not until the 20th Century did people begin to think that pacifism was relevant to an organized mass movement for social change. This 20th Century development reflects an inheritance from the 19th Century women's and labor movements—more people grasped analyses of the existing social system as unjust and violent and had experience in methods of mass organization, new varieties of direct action, and alternative institutions."

"*Colonial Peace Churches*
"As a result of religious conviction which rejected war, colonial peace churches asserted that the state has no authority over matters of conscience. This assertion brought them persecution by both European and colonial governments.

"Many of the peace churches, e.g., Mennonites, Brethren and Amish, took literally the New Testament injunction, 'Resist not evil.' Members came to be called *nonresistants* and, for the most part, followed a program of withdrawal from the world. These churches identified civil authority with the violence state governments employ to maintain their power. Therefore, they forbade *willful* participation in government, e.g., holding office or voting, since such action would bring members to occasions when they had to inflict injury or elect others who would inflict injury.

"Other colonial churches, e.g., the Society of Friends and Shakers, held a different world view. Instead of thinking of their churches as islands to be preserved from inundation by the world's sinfulness, they thought that their vision would overcome the world and they energetically promoted their ideas.

"Churches typified by the Quakers did not identify civil authority with state violence and so they did not object to active participation in politics. When the Society of Friends established the government of Pennsylvania, they intended that civil authority should flow from the power of the people's experience of 'Inner Light,' i.e., direct personal knowledge of the good."

"World War II COs had something else going for them that their World War I counterparts lacked: the beginning of a theoretical study of Gandhian nonviolence as a positive force for social change. Gandhi's brand of nonviolence emphasized building decentralized communities grounded in truth, justice and mutual aid, and encouraged the use of mass civil disobedience and noncooperation when the state interfered with the constructive program. Gandhi's work in India was popularized in the United States by Richard Gregg, A. J. Muste, Jessie Wallace Hughan, Reinhold Niebuhr, and others, and by the 1940s, pacifists had begun to implement American versions of Gandhi's program. They started with communities, variously called colonies or ashrams, and by the end of the war, they had gained more experience with organized direct action techniques."

—The Power of the People

How Can I Help? Stories and Reflections on Service
RAM DASS AND PAUL GORMAN
1985, Alfred A. Knopf

How Can I Help? is a practical and meditative inquiry into the nature of service. It grew out of the author's perception that many people are frustrated in their efforts to serve—and by *serve* they mean anything from being more loving with an ailing parent to making service your lifework—and that a guide exploring why that might be so would be of wide value. So the book itself is an act of service, and it embodies the spirit of compassion the authors seek to unleash in others.

The desire to ask, how can I help? is taken to be an essential quality of our being. In these times of feeling-impotent-in-the-face-of-planetary-peril, that affirmation alone has a healing effect. But the book ventures much deeper, into the psychological, cultural, and spiritual roots of the frustration of compassion. It turns out to be a paradoxical journey: To serve others is to continuously examine the self. In fact, self-awareness is the fountain from which compassion flows.

Most of the book is about how to keep the fountain flowing. To grapple with one's own (unavoidably ambiguous) feelings about suffering, loneliness, and death takes more than a little moxie, but it's the hard-but-true path to really being able to support others through those same experiences. If, as care givers, we're directly experiencing our own fears, we're better able to meaningfully listen to the needs of others.

How Can I Help? is not a purely philosophical book. In the process of their research, Ram Dass and Paul Gorman collected scores of personal testimonies from people engaged in service work; these stories, interspersed with the authors' text, provide real-life examples of the little confusions and revelations that daily strike those trying to help others. Their encouragement leaps off the pages.

I've found this to be a very useful book, so useful that I gave it to a friend who was considering going to Central America to volunteer for a peace organization, one who works with a woman with chronic fatigue syndrome, and another whose close friend was dying of AIDS. I dip into it often myself, because its wisdom is continuously renewing.

—JEANNE CARSTENSEN

"Being a long-term patient gives you a unique perspective on the world, and I have to laugh, although sometimes I must say it's a little bittersweet. . . . You'd be surprised at the number of people who talk to you and can't look you in the eye, even more than we normally can't look each other in the eye. It's like a parade of attitudes in here.

"It's funny; I laugh. I understand, I really do. I'm not a pretty picture. Their work is hard. But sometimes I just want to cry out, 'Hello! Is anybody there? Hello? Hello?' "

"We do not, however, insist that others go to the edge of their own pain. Under no circumstances do we judge or condemn another's suffering. We merely continue to work on our own, steadily, continuously, in all areas of our life. Opening to adversity and discovering in it all the places where we are clinging, resisting and denying, we gradually cut the cord between pain and suffering. The pain is there, but we can move beyond the suffering. To the extent that we ourselves are free of suffering, our very being becomes an environment in which others can be free of theirs if it is in the way of things."

"My idea was pretty simple at the beginning. I started to volunteer in wards with terminally ill children or burn victims—just go in there to cheer them up a little, spread around some giggles. Gradually, it developed that I was going to come in as a clown. . . . You see some pretty terrible things in these wards. Seeing children dying or mutilated is nothing most of us ever get prepared for. Nobody teaches us to face suffering in this society. We never talk about it until we get hit in the face.

"Like when I was starting out I was making the rounds one day at a children's hospital. The shade was pulled on this one room so I couldn't see, but I peeked in the door. It was a room with badly burned children in it. They had them in chrome crib beds with walls on the side, so they couldn't crawl out or fall out if it got too terrible in there.

"There was this one little black kid in one of them. He was horribly burned. He looked like burnt toast. Pieces of his face weren't there. Pieces of his ears were missing. Where was his mouth? You could hardly tell who he was. There was no way of pinning a person to the face, what little there was of it.

"It was just terrible, just mind-boggling. My jaw dropped, I gasped, and I came completely unglued. I remember flashing back to the antiwar movement. There was a picture of a napalmed kid I used to carry around at demonstrations. Suddenly here was that kid right in front of me. Unbelievably painful to behold.

"I was overwhelmed. And my mind went off in all sorts of directions. 'What's it going to be like if he lives?' 'What if I had a child this happened to?' 'What if this happened to me?'

"So there we were, burnt toast and unglued clown. Quite a sight, I bet. And I'm fighting just to stay there, trying to find a way to get past my horror.

"All of a sudden, this other little kid comes whizzing by—I think he was skating along with his IV pole—and he stops, and kinda pushes around me, and looks into the crib at this other kid, and comes out with, 'Hey. YOU UGLY!' Just like that. And the burnt kid made this gurgling laugh kind of noise and his face moved around, and all of a sudden I just went for his eyes, and we locked up right there, and everything else just dissolved. It was like going through a tunnel right to his heart. And all the burnt flesh disappeared, and I saw him from another place. We settled right in."

The Sacred: Ways of Knowledge, Sources of Life
PEGGY V. BECK AND ANNA L. WALTERS
1977, Navajo Community College Press*

Barney Mitchell, a Navajo traditional teacher, opens a chapter on shamanism saying, "The greatest sacred thing is knowing the order and structure of things." This respectful textbook explores the finest details of this "knowing," as expressed in scores of Native American sacred traditions from the Bering Sea to the deserts of the Southwest. Among the common threads: that learning can come from clowns, stories, and indirection; that everything is connected; that religion is a matter of daily mindfulness in the smallest of things; that wealth can be nonmaterial, and very different from greed. Hundreds of

* Navajo Community College Press, Tsaile (Navajo Nation), Arizona 86556

traditional teachers and participants—Navajo, Seminole, Tewa, Iglulik Eskimo—are quoted, from anthropological reports and direct interviews. Their words live, not as an archaic and romantic remnant, but as badly needed wisdom that could help us connect with the sources of life, rather than destroy them.

There's a sad undercurrent to this book, as it describes the shock of conquest and whites' attempts to suppress Native American "giveaways," dances, ball games, and religions from the 1870s until the 1920s. Children were packed off to mission schools and prohibited from speaking their own languages. Conquest and forced conversion prompted some Native Americans to develop syncretistic religious responses, like the Ghost Dance, Peyotism, and Bole Maru. Other Native Americans tried to integrate hunting rifles, snowmobiles, and Christianity into traditional cultures and rituals. Some simply forgot—in *The Sacred,* songs are quoted that nobody sings any more. A few peoples—especially the Pueblo cultures—remained economically independent and preserved traditional lives intact. This textbook, designed for Native American community college students, provides encouraging examples of successfully preserving traditional sacred life within the dominant culture. For example, after the birth of her first baby, one college-educated Seminole woman kept detailed taboos, eating and sleeping separately from her husband even though they shared a house trailer in the town where her husband worked as a school teacher. Instead of a vague romance about Indian ways, this book provides a depth of details.

—KATY BUTLER

"In Chapter One, part II, we discussed certain concepts of the sacred that many Native American tribal people share. Among these concepts were:

1. A belief in or knowledge of unseen powers: The Great Mystery or the Mysteries.
2. Knowledge that all things in the universe are dependent on each other.

"Reflected in many sacred systems of knowledge is the idea that the world is a mystery and yet, at the same time, the world operates according to strict, ordered relationships. Some questions can never be answered with a single answer and others can be answered simply by observing various cycles day after day, year after year; cycles like day and night, the changing seasons, and the cycles of growth in humans, plants, and animals."

"Lame Deer sums up the special relationship of Clowns to sacred power when he says: 'To us a clown is somebody sacred, funny, powerful, ridiculous, holy, shameful, visionary. He is all this and then some more. Fooling around, a clown is really performing a spiritual ceremony. He has a power. It comes from the thunder-beings, not the animals or the earth. In our Indian belief, a clown has more power than the atom bomb. This power could blow off the dome of the Capitol. I have told you that I once worked as a rodeo clown. This was almost like doing spiritual work. Being a clown, for me, *came close to being a medicine man. It was in the same nature.'* (Lame Deer, 1972:236.)"

Walden
HENRY DAVID THOREAU
1854; 1942; 1960, Houghton Mifflin Company

Walden may be, as E. B. White has said, the most idiosyncratic book in American literature. It is also perhaps the most outrageous. Although *Walden* rightfully occupies a central position in the canon of essential environmental texts, it resists such categorization. It is thoroughly anarchistic, constantly encouraging its reader to question any authority—even that of its author—which would undermine the integrity of the individual. Yet it does not encourage the sort of "rugged individualism" that is often associated with both the American frontier and American business; instead, *Walden* invites in its reader the development of a highly individuated personality, one that recognizes and celebrates the great web of being, the fact that everything is indeed connected to everything else. In short, it is one of the finest expressions of what we would now call an ecological attitude.

The interested reader, however, should not stop here. In *The Maine Woods* and his copious *journals*, Thoreau presents the eloquent testimony of one who thinks globally while observing and acting locally. Three of his shorter essays—"Walking," "Autumnal Tints," and "Wild Apples"—are of special interest to those pursuing a philosophy of the wild, and his "Civil Disobedience" has been used as a handbook by environmental activists.

Keep in mind that, although Thoreau urges his readers, "Simplify! Simplify!," his writings are anything but simplistic. His work is full of the mystery that surrounds us all the time. "Talk of mysteries," he exclaims in *The Maine Woods*, "Think of our life in nature,—daily to be shown matter, to come into contact with it,—rocks, trees, wind on our cheeks! the *solid* earth! the *actual* world! the *common sense*! *Contact*! *Contact*! *Who* are we? *where* are we?"

—Sean O'Grady

"The surface of the earth is soft and impressible by the feet of men, and so with the paths which the mind travels. How worn and dusty, then, must be the highways of the world, how deep the ruts of tradition and conformity."

"We might try our lives by a thousand simple tests; as, for instance, that the same sun which ripens my beans illumines at once a system of earths like ours. If I had remembered this it would have prevented some mistakes. This was not the same light in which I hoed them. The stars are the apexes of what wonderful triangles! What distant and different beings in the various mansions of the universe are contemplating the same one at the same moment!"

"There is an incessant flux of novelty into the world and yet we tolerate incredible dullness. I need only suggest what kind of sermons are listened to in the most enlightened countries. There are such words as joy and sorrow, but they are only the burden of a psalm, sung with a nasal twang, while we believe in the ordinary and mean."

"The life in us is like the water in the river. It may rise this year higher than man has ever known it, and flood the parched uplands; even this may be the eventful year, which will drown out all our muskrats. It was not always dry land where we dwell. I see far inland the banks where the stream anciently washed, before science began to record its freshets."

Tao Te Ching
Lao Tsu, translated by Gia-fu Feng and Jane English
1972, Vintage Books

FIFTY-ONE

All things arise from Tao.
They are nourished by Virtue.
They are formed from matter.
They are shaped by environment.
Thus the ten thousand things all respect Tao and honor Virtue.
Respect of Tao and honor of Virtue are not demanded,
But they are in the nature of things.

Therefore all things arise from Tao.
By Virtue they are nourished,
Developed, cared for,
Sheltered, comforted,
Grown, and protected.
Creating without claiming,
Doing without taking credit,
Guiding without interfering,
This is Primal Virtue.

Further Reading

There are over a hundred good books—some practical, some intellectual, some purely aesthetic, and some philosophical—glossed in this reading list. Many hundreds more lie beyond our reach or grasp. The selections of nature writing, for instance, only hint at the vast and beautiful literature that celebrates the living world. The books listed here, though, may make you want to trace their tributaries, and pique your curiosity about other realms of knowledge, too. The list is idiosyncratic, skewed to the humanities, reflecting the reading preferences of unreconstructed liberal arts majors. Not aspiring to be as comprehensive as a bibliography, this is just to suggest some further reading.

BACHELARD, GASTON, *The Psychoanalysis of Fire*, Beacon Press, 1964. A prolonged meditation on the relation of the human psyche to flame.

BALDWIN, J., ed., *The Essential Whole Earth Catalog*, Doubleday, 1986. The latest in a long and happy line of *Whole Earth Catalog* compendia which provide, through provocative, astute reviewing, "access to tools"—books, primarily—to implement the proposition that "we are as gods and might as well get good at it."

BATES, MARSTON, *The Forest and the Sea: A Look at the Economy of Nature and the Ecology of Man*, 1960; Lyons and Burford,

209

1988. Deriving ecological principles from similarities of patterns in rain forests and coral reefs, Bates writes an artful introduction to his science.

BATESON, MARY CATHERINE, *With a Daughter's Eye: A Memoir of Margaret Mead and Gregory Bateson,* Pocket Books, 1985. This author bears some of the most enviable DNA in the world, worthily. With a remarkable blend of objectivity and intuition, Mary Catherine Bateson describes her parents, two very human individuals, both possessed of towering intellect, care for the living world, and original vision. Mead, Bateson, and Bateson (M. C.) are all figures whose work is essential to know. (For more on Gregory Bateson, see the entry under David Lipset.)

BERGER, JOHN, ed./Restoring the Earth Project, *Environmental Restoration: Science and Strategies for Restoring the Earth,* Island Press, 1990. Long-awaited guidance for serious environmental restoration projects. This book is the result of a four-day ground-breaking conference on ecological restoration held at the University of California, Berkeley, to consider the restoration of all major natural resource systems and the planning of environmentally suitable urban areas; the book provides an introduction and an overview of restoration work for the professional, scientific, and academic communities. It also presents examples of successful projects and gives case study guidelines for when and how to use restoration techniques.

BERMAN, MORRIS, *The Reenchantment of the World,* Bantam Books, 1984. A theory about how we lost the sense of sacred participation with nature that our hunter-gatherer ancestors experienced.

BESTON, HENRY, *The Outermost House,* 1928; Penguin Books, 1976. An account of a solitary year spent on Cape Cod, permeated with Beston's perception of the relation of humans to the cosmos. This slender volume, which first appeared in 1928, was written in longhand on the kitchen table in a little room with windows overlooking dunes and the North Atlantic. Beston describes with rhythmic, lyric intensity the wonder and mystery he beheld out his windows.

BLUMBERG, LOUIS AND ROBERT GOTTLIEB, *War on Waste: Can America Win Its Battle with Garbage?*, Island Press, 1989. A hard-hitting, provocative account of the history and politics of waste production and management in the United States. The centerpiece of the book is the juicy saga of the Los Angeles City Energy Recovery Project (LANCER), a massive waste incineration scheme killed by community opposition from both environmentalists and low-income neighborhoods.

BOOKCHIN, MURRAY, *Our Synthetic Environment*, Alfred A. Knopf (out of print), 1962. This book preceded *Silent Spring* and has the particular strength of being a good early critique of industrial civilization.

BRENNEMAN, RUSSELL L. and SARAH M. BATES, eds., *Land-Saving Action*, Island Press, 1984. A manual on private-sector preservation of open space written by twenty-nine experts in the field. It explains how to form a land trust; criteria for agricultural, community, and urban land trusts; how to manage and protect them; and many other nuts-and-bolts land-saving strategies.

BROOKS, PAUL, *Roadless Area*, Alfred A. Knopf (out of print), 1964. The author's own journeys through wilderness areas. This book conveys the joy of exploration, while making a clear statement about the need to preserve North American wild places. A Burroughs Medal winner.

BROWN, TOM, JR. and BRANDT MORGAN, *Tom Brown's Field Guide to Nature Observation and Tracking*, Berkley Publishing Group, 1983. A straightforward, plainspoken, useful handbook for developing the extraordinary state of awareness required for reading the narrative of nature, following the story of a wild thing's passage.

BUBER, MARTIN, et al., *I and Thou*, 1958; Charles Scribner's Sons, 1970. A highly influential book about mutuality; about the difference between experience of the world and relation with it. An eminent Jewish philosopher's early twentieth century discourse. It is conducive to a vital, reverent connection with a larger reality.

211

CARRIGHAR, SALLY, *One Day on Beetle Rock,* 1944; University of Nebraska Press, 1978; *One Day at Teton Marsh,* 1947; University of Nebraska Press, 1978; *Home to the Wilderness: A Personal Journey,* Houghton Mifflin (out of print), 1973. Carrighar was early (1940s) to visit the wilderness and to recount its dramas with flair and accuracy. Gripping, intricate tales. Her "personal journey" explains the childhood sorrow out of which she emerged into nature as home.

CEIP FUND, *The Complete Guide to Environmental Careers,* Island Press, 1989. Nuts-and-bolts information for the recent graduate aspiring to right livelihood in the eco-biz. Up-to-the-minute, the *Complete Guide* analyzes environmental employment prospects in various sectors of the public and private sector and supplies practical advice on first choosing, then entering, a particular field.

CROSBY, ALFRED W., *Ecological Imperialism: The Biological Expansion of Europe 900–1900,* Cambridge University Press, 1986. Exotic species brought to America from Europe greatly changed the ecology here.

DALY, HERMAN E., ed., *Economics, Ecology, and Ethics: Essays Toward a Steady-State Economy,* W. H. Freeman, 1980. Herman Daly is a thoughtful, literate observer of the planetary problem (growth) and a real-live economist to boot. In his contributions to this anthology, Daly uses the lens of economics with uncommon realism. His fellow contributors, Hardin, Boulding, Schumacher, Ophuls, Randers, and Meadows, are eminently capable of reasoning paths toward an economy that would exhibit the same stability as an undisturbed ecosystem. A challenging but essential ethical and theoretical study.

DAVIS, DONALD EDWARD, *Ecophilosophy: A Field Guide to the Literature,* R. & E. Miles, 1989. This book cuts a wide swath through eco-philosophy, from traditional philosophers to modern thinkers working in the various disciplines that are all contributing to the evolution of eco-philosophical thought: human ecology, animal rights, ecological feminism, theology, ecology, and others. The annotations are long enough that you'll know whether or not you want to bother actually reading the book.

DE ANGULO, JAIME, *Indian Tales,* Farrar, Straus & Giroux, 1962. Ostensibly a children's book, this account of a family's (Bear, Antelope, Fox, and Quail) journey from the interior mountains of California to the coast, with encounters with elders and storytellers along the way, warmly conjures the experience of a land-based traditional people.

DEVALL, BILL and GEORGE SESSIONS, *Deep Ecology: Living As If Nature Mattered,* Gibbs Smith, 1987. A fundamental exposition of the philosophy of deep ecology, which dethrones *Homo sapiens* and is premised on the value that all life is equal.

DEVOTO, BERNARD, ed., *The Journals of Lewis and Clark,* Houghton Mifflin, 1973. Hands down the most readable version of Lewis and Clark's journals of their legendary expedition in 1804 to 1806. With excellent commentary by historian DeVoto.

DILLARD, ANNIE, *Pilgrim at Tinker Creek,* 1974; Harper & Row, 1988; *Teaching a Stone to Talk: Expeditions and Encounters,* Harper & Row, 1982. Dillard is one of our most brilliant writers, bar none. She looks at things truly and is a master at conveying both the luminosity and unsentimentality of nature.

DURELL, LEE, *State of the Ark: An Atlas of Conservation in Action,* Doubleday, 1986. The state of the ark is alarming, as portrayed in another studded-with-visuals book in the Gaia series. It is closely focused on ecosystems and particular species, the often grisly perils they face, and the hopeful, doughty bands of citizen-defenders coming to the rescue. (See Norman Meyers' *Atlas of Planet Management* for more.)

ECHEVERRIA, JOHN D., POPE BARROW, and RICHARD ROOS-COLLINS, *Rivers at Risk: The Concerned Citizen's Guide to Hydropower,* Island Press, 1989. This timely hydropower handbook gives citizens a practical understanding of how to influence government decisions on proposals for hydropower development of the nation's rivers. It also serves as a vital resource for attorneys, engineers, and professional conservationists involved with or interested in hydropower issues.

EHRLICH, GRETEL, *The Solace of Open Spaces,* Penguin Books, 1985. Wyoming from the perspective of a woman who moved there to mend a broken heart. Splendid writing lauds the places truly, if slightly romanticizing the people.

EHRLICH, PAUL R. and JOHN P. HOLDREN, eds., *The Cassandra Conference: Resources and the Human Predicament,* Texas A & M University Press, 1988. A generally gloomy update on the big grim picture, with the virtue of providing the current opinions (and work) of some of the preeminent scientist-doomsayers. Fine perspective-shifting conference envoi by Donella Meadows.

EISELEY, LOREN, *The Star Thrower,* Harcourt Brace Jovanovich, 1979. This surpassing volume is Eiseley's own selection from his prose and poetry. A brooding literary genius, anthropologist, historian of science, Eiseley writes piercingly of evolution, fox kits, and starfish—and of the Earth in its genesis and the prospect of humanity. Stunning eloquence; deep, and often nocturnally dark, thought.

EMERSON, RALPH WALDO, *Nature,* 1836; Beacon Press, 1985. A founding text in much of American nature writing.

FAULKNER, EDWARD H., *Plowman's Folly and A Second Look,* 1943; Island Press, 1987. A pioneering critique, followed by a qualifying volume, of the use of the moldboard plow in agriculture; an indictment of said plow as the leading culprit in the erosion and depletion of our nation's soils, and a detailed report of the author's homegrown research on an alternative method—conservation tillage.

FAULKNER, WILLIAM, *The Bear, Three Famous Short Novels: Including Spotted Horses; Old Man; Bear,* Random House, 1958. A multileveled novel: An anthem to the end of wilderness, this story about hunting and the end of innocence uses the central motif of the hunt to explore the relationship to wild nature of a culture (southern) that was in the process of becoming urbanized.

FLANAGAN, JOAN, *The Grass Roots Fundraising Book: How to Raise Money in Your Community,* Contemporary Books, 1982. This book addresses the subject of its title, and more. It counsels both common sense and integrity for the organizations and leaders struggling to succeed in their aims, which may or may not require fund-raising.

FLETCHER, COLIN, *The Complete Walker III,* Alfred A. Knopf, 1984. It is widely agreed to be the best handbook for hikers, period. The book is engagingly written and trustworthily opinionated. The volume to consult before you head for the hills (or deserts or canyons).

FOREMAN, DAVE and HOWIE WOLKE, *The Big Outside,* Ned Ludd Books, 1989. Earth First! cofounders Dave Foreman and Howie Wolke's inventory of big wilderness areas left in the lower forty-eight. Every roadless area of a hundred thousand acres or larger in the western states or fifty thousand or larger in the eastern states is listed and described. Lots of plain speech about the major threats to each area.

FORSYTH, ADRIAN and KEN MIYATA, *Tropical Nature: Life and Death in the Rain Forests of Central and South America,* Charles Scribner's Sons, 1984. Naturalists writing knowledgeably and with enthusiasm about the wonders and complexities of tropical rain forest life forms. You want creepy-crawlies and jungle rot? We got 'em. And, what's more, we love and are coming to understand them, the authors say.

FRANCIS, MARK, LISA CASHDAN, and LYNN PAXSON, *Community Open Spaces: Greening Neighborhoods Through Community Action and Land Conservation,* Island Press, 1984. By documenting the evolution of several community parks and gardens in New York City, this book can introduce architects, planners, and neighborhood activists to the option of community-maintained open space.

GOLDSMITH, EDWARD and NICHOLAS HILDYARD, eds., *The Earth Report: The Essential Guide in Global Ecological Issues,* Price/Stern/Sloan, 1988. Worldwatch without the affection for the nation-state. Penetrating articles on several major topics of global concern plus an encyclopedia of ecological trends, issues, and disaster, all treated with the punch that only Teddy Goldsmith and his *Ecologist* gang can deliver.

GRADWOHL, JUDITH and RUSSELL GREENBERG, *Saving the Tropical Forests,* Island Press, 1988. "Experience in, or insights into, methods by which [tropical] forests could be preserved, in varying states from pristine to disturbed, in ways consistent with continuing economic advance for tropical peoples." Hopeful and fascinating case studies in the establishment of forest reserves, sustainable agricultural practices, natural forest management, and tropical forest restoration projects.

GRAY, ELIZABETH DODSON, *Green Paradise Lost,* Roundtable Press, 1979. Formerly titled *Why the Green Nigger?,* this book contributed an early analysis of our civilization's hierarchical ranking of, and preference for, different beings, with white men at the top of the pyramid, and vegetation and soil at the very bottom, being treated accordingly.

HAINES, JOHN, *The Stars, the Snow, the Fire,* Graywolf Press, 1989. In 1947, Haines moved north of Fairbanks to homestead. This book of his essays evokes the landscape and life of the Far North. Personal, poetic, and direct, Haines, a trapper, talks about hunting. A haunting book, much concerned with death and blood.

HAY, JOHN, *In Defense of Nature,* Little, Brown, 1969. Collected reflections of a contemporary writer on living creatures and nature's cycles. Hay's conviction is that nature alone can bring us back to sanity.

HAYS, SAMUEL P., *Conservation and the Gospel of Efficiency: The Progressive Conservation Movement, 1890–1920,* Harvard University Press, 1986. A heavily researched, exhaustive historical analysis of the real politics of conservation in the early

216

twentieth century, which was, to an extent, a struggle between government technocrats (like Gifford Pinchot) and local interests—sometimes developers, sometimes preservationists—to determine the fate of U.S. public lands. *Conservation* continues to mean very different things to its various standard-bearers, and this history details the early politics of those differences.

HENDERSON, HAZEL, *The Politics of the Solar Age: Alternatives to Economics,* 1981; Knowledge Systems, 1988. Hazel is a buoyant, iconoclastic world-mom, confidently breezing into all manner of expertise sanctuaries to announce that the paradigm (economic, diplomatic, medical, sexual, political) is different now, that systems are integrated now, that winning coalitions of former losers are possible now, that, in short, the wool is no longer securely over the eyes. She synthesizes the new conceptual developments of our era and covers global citizen activism with an adamantly positive attitude.

HUNT, CONSTANCE ELIZABETH with VERNE HUSER/NATIONAL WILDLIFE FOUNDATION, *Down by the River: The Impact of Federal Water Projects and Policies on Biological Diversity,* Island Press, 1988. Watershed death knells, tax-dollar financed. This careful description of which species and associations bite the dust as a result of Uncle Sam's reclamation and channelization projects shows causes and effects plainly; also recommends commonsense (always scarce within scent of a pork barrel) policy reforms.

HYNES, H. PATRICIA, *The Recurring Silent Spring,* Pergamon Press, 1989. A perceptive, unflinching eco-feminist analysis of the significance of *Silent Spring,* the meaning of Rachel Carson's life, and the ultimate ineffectuality of the federal bureaucracy's response to the root problem—the presumption that nature exists for the use of man, which Carson decried. An original, necessary, and deeply thought-provoking book.

JOHNSON, WARREN, *Muddling Toward Frugality,* Sierra Club Books, 1978. Johnson argues for slow-but-steady transformation from an industrial to an ecological society.

KOHR, LEOPOLD, *The Breakdown of Nations* (out of print); *Development Without Aid: The Translucent Society,* Schocken Books (out of print), 1979; *The Overdeveloped Nations: The Diseconomies of Scale,* Schocken Books (out of print), 1978. Kohr was a mentor to E. F. Schumacher and, having outlived his friend, continues from the podium to elucidate the follies and futility of gigantism in the nation-state, the theme of his books. Appropriate scale for effective governance is among Kohr's prime concerns. A steadfast, practical thinker, Kohr embodies and values joie de vivre as a quality possible to human settlements, given a reasonable politics. With the breakdown of governments in Eastern Europe, his works seem remarkably prescient.

KROEBER, THEODORA, *Ishi in Two Worlds: A Biography of the Last Wild Indian in North America,* 1961; University of California Press, 1976. The subtitle tells it all. The lone survivor of the Yahi Indians of California, Ishi was a wholly dignified human being whose emergence from the Stone Age to the modernity of 1911 fundamentally challenged people's preconceptions of "primitive" and "Indian."

KRUTCH, JOSEPH WOOD, *The Best Nature Writing of Joseph Wood Krutch,* William Morrow, 1969; *The Great Chain of Life,* Houghton Mifflin, 1957. Erudite, humane, graceful considerations of evolution and observations of nature. Rich with implication.

LEGUIN, URSULA, *Always Coming Home,* Bantam Books, 1987. Almost an anthropological account (including details like the songs and diet) of the Kesh, a future people dwelling on the Pacific coast. What's wonderful about the Kesh is that they're a people who live very close to the Earth. The original edition came with a cassette of Kesh music.

LIPSET, DAVID, *Gregory Bateson: The Legacy of a Scientist,* Beacon Press, 1980. A "Child's Garden of Bateson," whose own books, *Steps to an Ecology of Mind* and *Mind and Nature,* are a mite demanding, however eager the layperson reading them. Bateson worked interestingly and often brilliantly in anthropology, ecology, ethology, psychology, cybernetics, and evolutionary biology, virtually all the sciences which hold the potential to inform a sustainable culture. Ergo, access to Bateson is access to realms of thought which are crucial. Lipset could move a reader to a point of entry to works by the man himself.

LITTLE, CHARLES E., *Green Fields Forever: The Conservation Tillage Revolution in America,* Island Press, 1987. A brisk and perceptive account of a plowless form of mechanized, chemical agriculture that reduces water and wind erosion and promotes soil health by leaving crop residues on field surfaces.

LOEFFLER, JACK, *Headed Upstream: Interviews with Iconoclasts,* Harbinger Books, 1989. Other than the fact that Loeffler includes only one woman in this tasty group of fourteen iconoclasts (which is hardly the ratio in which contrary females actually occur), he's thrown a very interesting party-on-the-page, holding conversations with the likes of Ed Abbey, Dave Foreman, Garrett Hardin, Doug Peacock, and Gary Snyder. Loeffler's own pieces on the assault on the Navajo and Hopi peoples' Four Corners homeland are included as well. Good entrée to avant-garde intelligence on fundamental issues.

LONDON, JACK, *The Call of the Wild,* 1903; Vintage, 1989. Mighty gripping dog story. Buck, the hero-dog, is pressed into servitude, suffers, and finally attains freedom. It could get you looking at your pets, indeed all beasts, differently. And London's writing is a national treasure. It glories in the rawness of nature, and vanquishing it. A romanticized portrait of the Wild.

LOVINS, AMORY and HUNTER LOVINS, *Brittle Power,* Brick House, 1983. America's foremost soft-path energy mavens meticulously analyze the United States' dependence on centralized power generation and attenuated lines of supply and find the system to be, among other things, a national security risk amenable to quick disruption. One well-placed bomb could black out a seaboard.

LOWRY, SUSAN MEEKER, *Economics As If the Earth Really Mattered,* New Society, 1988. A sensible guide to socially responsible investing, not only in ethical corporations, but in institutions that contribute to the health of the community. You don't have to be rich to use this volume.

MANDER, JERRY, *Four Arguments for the Elimination of Television,* William Morrow, 1978. Not quite Ludditism, but a radical critique of this technology and its perversion of our understanding of the natural world and of reality in general.

MANSBRIDGE, JANE J., *Beyond Adversary Democracy,* University of Chicago Press, 1980. A sociologist's insightful, observant case for the existence of two types of democratic processes: adversary, which may serve to reconcile conflicting interests, and unitary, which assumes common interests and entails different processes. It is cited here for its potential value to grassroots activists. A social scientist's description of what works and how, in a New England town meeting and for a help line collective. Operating under assumptions appropriate to the particular type of democracy—adversary or unitary—enhances the effectiveness of each.

MANTELL, MICHAEL A., STEPHEN F. HARPER, and LUTHER PROPST/ THE CONSERVATION FOUNDATION, *Creating Successful Communities: A Guidebook to Growth Management Strategies,* Island Press, 1989. A how-to compendium of techniques for effective land use and growth management to help communities retain their individuality in the face of rapid growth. The book includes a framework for land-use decision making and growth management; techniques for protecting key resources such as agricultural land, open space, historic and cultural structures, aesthetics, and rivers and wetlands, as well as ways to effectively organize for action. Also available is a workbook, *Resource Guide for Creating Successful Communities.*

MARSTON, ED, ed., *Reopening the Western Frontier: From "High Country News,"* Island Press, 1989. An anthology of well-written pieces illuminating the realities of the West today. Culturally sensitive treatments of the tensions and destructions brought about in specific places throughout the current boom and bust of resource exploitation (also known as development, depending on who profits).

MASER, CHRIS, *The Redesigned Forest*, R. & E. Miles, 1988. "I see no sustainable forests . . . because humanity is summarily cutting them down, and, where possible, replacing forests with fast-growing tree plantations." Starting from there, research biologist Maser weaves an eco-philosophical inquiry into the ethical, psychological, and biological bases for a practice of sustainable forestry. A wise, compassionate, holistic statement.

MATTHIESSEN, PETER, *Wildlife in America*, Viking Press, 1959; *Indian Country*, Viking Press, 1984; *The Snow Leopard*, Viking Press, 1978; *The Tree Where Man Was Born*, E. P. Dutton, 1972. Both a naturalist and an important literary figure, Peter Matthiessen produces work that is characterized by precise observation and salient detail. A courageous traveler, Matthiessen ventures widely about Earth—from the Himalayas to East Africa—and deeply within himself.

MCHARG, IAN L., *Design with Nature*, Natural History Press, 1971. If society holds together long enough to afford us the leisure of good planning, this will be an ur-text.

MCINTYRE, JOAN, *Mind in the Waters: A Book to Celebrate the Consciousness of Whales and Dolphins*, Charles Scribner's Sons (out of print), 1974. Brilliant anthology that invites the reader to consider the possibility of other, equal consciousness being on the planet, and in the province of Cetacea—great whales and dolphins, singing, playing, perhaps even displaying wisdom.

MCNEILL, WILLIAM H., *Plagues and Peoples*, Doubleday, 1977. A survey of human history as affected by human coevolution with and adaptation to pathogens and epidemic disease. Interesting description of cultural attitudes toward leprosy, for instance.

MELLO, ROBERT A., *Last Stand of the Red Spruce*, Island Press, 1987. A carefully buttressed and persuasively argued case indicting air pollution generally, and acid rain specifically, as the culprit in the dying off of the red spruce, a tree species native to the boreal regions of northeastern North America. Mello describes the history of the research on the spruce forests' decline, the scientific controversies raised and resolved, and the inadequacy of government's response to date.

MERCHANT, CAROLYN, *The Death of Nature: Women, Ecology, and the Scientific Revolution*, Harper & Row, 1983. Just what it says. A history of the growth of the attitude that the Earth is inanimate and without spirit or feeling, an attitude which, by no coincidence, accompanied the burgeoning of extractive industries and industrialism. That Earth was a person was once a widespread belief through the West, not just a quaintness of tribal peoples.

MEYERS, NORMAN, *Gaia: An Atlas of Planet Management*, Doubleday, 1984. The notion of planet management reeks of hubris, but the book itself is a richly graphic, politically savvy atlas of the problems of the global environment, their impacts on the human, and other, species, and some of the positive responses being made around the world. It is useful and quite cosmopolitan, long on specifics, thus a handy volume for introducing the realities of the global ecological crisis. (See Lee Durell's *State of the Ark* for more.)

MILLER, G. TYLER, JR., *Living in the Environment: An Introduction to Environmental Science* (fifth edition), Wadsworth, 1988. A big, useful reference book. A college text, actually, with absolutely clear explanations of all major biological systems, mechanics of pollution sources and alternative technologies, maps, glossaries, appendices, pie charts, sidebars for controversies. A shield and a buckler for the activist who wants to go informed.

MOLL, GARY and SARA EBENRECK, eds./AMERICAN FORESTRY ASSOCIATION, *Shading Our Cities: A Resource Guide for Urban and Community Forests*, Island Press, 1989. A comprehensive handbook for citizen activists or public officials intent on greening their communities and/or caring for and expanding existing urban tree plantings. Good information about city subculture, legal issues, education, and organizing approaches. Upbeat and handy.

MOORE, ANDREW OWENS, *Making Polluters Pay: A Citizen's Guide to Legal Action and Organizing,* Environmental Action Foundation, 1987. Suing the bastards isn't necessarily the most efficient way to bring corporate poisoners to justice, and no amount of money can compensate an innocent body deformed or terminally sickened by toxics in the water supply, but when deciding whether to go to court, citizens will find this a clear and handy guide.

MOWAT, FARLEY, *Never Cry Wolf,* Little, Brown, 1963; *Woman in the Mists,* Warner Books, 1987. A Canadian writer of passionate, spellbinding, and sometimes funny (as in *Never Cry Wolf*) tales in defense of wildlife. Mowat's *Never Cry Wolf* was one of the first books to show the other side of wolves. *Woman in the Mists* is fierce with understanding of the bonds Dian Fossey formed with "her" gorillas. For some reason the Reagan administration deemed Mowat to be dangerous for us to know about (perhaps for his antibureaucratic stance) and barred him from entry to the United States.

NABHAN, GARY PAUL, *The Desert Smells Like Rain,* North Point Press, 1982; *Gathering the Desert,* The University of Arizona Press, 1985; *Enduring Seeds,* North Point Press, 1989. Nabhan is an ethnobotanist who concerns himself with the indigenous plants and peoples—the cultures and agriculture—of the Sonoran bioregion, primarily. It is a wild terrain, rich in moral implication. Nabhan, a John Burroughs medalist, writes superbly; widely learned, he communicates vital information with accurate grace.

NASH, RODERICK, *Wilderness and the American Mind,* Yale University Press, 1982. The classic study of the evolution of our attitudes about wilderness. Nash traces its development back to primitive times, but his primary emphasis is on American attitudes, from an era of regarding the whole continent as wilderness to the dawning realization that the frontier was closed. Nash's rigorous chef d'oeuvre illuminates our very complicated relationship to wilderness and has been updated several times.

NELSON, RICHARD K., *Make Prayers to the Raven,* University of Chicago Press, 1986; *The Island Within,* North Point Press, 1989. A blend of ethnography, personal reflections, and natural history, *Make Prayers to the Raven* illuminates the Koyukon view of the environment, by a cultural anthropologist who happens to write exquisitely clean prose. *The Island Within,* a masterwork, is the living story of Nelson's awe-filled, intimate learning of an island hunting ground off his south-coast-Alaska home.

NEWSDAY, *Rush to Burn: Solving America's Garbage Crisis?,* Island Press, 1989. This provocative treatment of the garbage crisis in the United States, with a critical look at the incineration techno-fix, is a compilation of articles from Long Island's *Newsday.* *Rush to Burn* describes the politics, the pros, and mostly, the cons of putting our wastes to the flames.

NIEHARDT, JOHN C., *Black Elk Speaks,* Washington Square Press, 1932. Oglala Sioux shaman Black Elk lived through the desperate times of the late nineteenth century that ended with the massacre at Wounded Knee, South Dakota, on 29 December 1890. In 1930 he decided to tell his life story to John Niehardt, who made it into this beautiful and widely read book. At its heart is Black Elk's "power-vision," which came to him at nine years old, a magical tale detailing the Sioux vision of the interconnectedness of all life and prophesying the downfall of the Sioux nation. The vision reveals truths for all peoples, not the least of which could be our own.

O'TOOLE, RANDAL, *Reforming the Forest Service,* Island Press, 1988. A cogent analysis of the proximate cause of the Forest Service's devastating management practices (that being federal subsidy of the agency's below-cost timber-selling activity) and a sweeping proposal to reform this practice by invoking the market to rationalize, through pricing mechanisms, the use of all the values of our forest, including recreation.

PADDOCK, JOE, NANCY PADDOCK, and CAROL BLY, *Soil and Survival: Land Stewardship and the Future of American Agriculture,* Sierra Club Books, 1986. "We must ask," write the authors,

"what ideas and actions make up the Ethical approach to farmland that will produce good feedback, good karma for us in
coming generations." And so they do, in an exploration of the
once and future psychology of a mutually nurturant relationship between humanity and cultivable land; a consideration in
which they boldly range from straightforward portrayal of the
present reality of American farmers through comparative religion and poetry; a remarkably insightful work, making substance out of the dialectic between urban and rural culture.

PALMER, TIM, *Endangered Rivers and the Conservation Movement,*
University of California Press, 1986. A comprehensive, deeply
felt, and exhaustively researched history of river conservation in
the United States. An indispensable work on this critical seg-
ment of the preservationist cause.

PALMER, TIM, *The Sierra Nevada: A Mountain Journey,* Island Press,
1988. A natural history of the Sierra that brings the range, the
people, and the communities to life through penetrating descriptions of people, thrilling adventures, and fine photography.
The story of the Sierra Nevada's development, from the gold
rush days to the modern battles of the 1980s and the struggle for
the future.

PAWLICK, THOMAS, *A Killing Rain: The Global Threat of Acid Precipitation,* Sierra Club Books, 1984. Thorough, graphic, information rich, and remarkably readable, this book covers the
entire content of the subject of acid rain, from denuded landscapes to the hope for cleaner, sweeter skies implicit in diver-
sified, decentralized, efficient energy-generation technology.

PEATTIE, DONALD CULROSS, *A Natural History of Trees of Eastern
and Central North America,* Bonanza Books (out of print),
1964. A gorgeous encyclopedia of America's tree species; lav-
ishly written descriptions. Sensual accounts of the sight, smell,
and sound of different trees, along with anecdotes about such
things as what their respective woods were best for.

PITTMAN, NANCY P., ed., *From The Land,* Island Press, 1988. A
wonderful anthology of articles from *The Land,* an extraordi-

nary magazine sprung from the soil, which flourished from 1941 to 1954. Henry Wallace, Louis Bromfield, Ralph Borsodi, Gifford Pinchot, Aldo Leopold, Paul Sears, Alan Paton, and Wallace Stegner were among the more illustrious contributors to *The Land,* and their works were interspersed with equally heartfelt and original pieces from plain citizens of rural America. A little-known cultural treasure.

PLANT, JUDITH, ed., *Healing the Wounds,* New Society, 1989. An anthology of writings by women on not just ecology, but the whole system of oppression that is shattering Earth, culture, family, and psyche. Bluntly unapologetic critiques of the dominant paradigm. Women stepping out to take on the big responsibility.

PLATT, RUTHERFORD, *The Great American Forest,* Prentice Hall (out of print), 1965. An astonishing amount of information about the evolution of the American silva, its history, and physiology. An interesting grab bag with a mix of categories, from genera to anecdotes of wilderness exploration to portrayal of forest as macro-organism.

QUAMMEN, DAVID, *Natural Acts: A Sidelong View of Science and Nature,* Schocken Books, 1985. A columnist for *Outside* magazine, Quammen is a first-rate essayist who popularizes scientific information with a great sense of humor and a nice turn of phrase.

RAPHAEL, RAY, *Tree Talk: The People and Politics of Timber,* Island Press, 1981. A vivid and encompassing depiction, verbatim and with illustrations, of the timber economy of the West. The emphasis is definitely on economics, and Raphael is for creative reform of forestry rather than for absolute forest preservation, which may or may not be sane realism. But for the environmentalist whose notions of forestry, logging, and its perpetrators are all vague, Raphael's book intelligently renders the pictures and the sound track of a big, neobiological industry. Good ear and good illustrations.

REISNER, MARC and SARAH BATES, *Overtapped Oasis: Reform or Revolution for Western Water,* Island Press, 1989. A critique of the cardinal dogma of the American West: that the region is always running out of water and must therefore build more and more dams. The book analyzes the West's water allocation system from top to bottom—at the state, federal, and local or regional levels—and offers dozens of revolutionary proposals for increased efficiency and policy reform.

RICHTER, CONRAD, *The Fields,* Alfred A. Knopf (out of print), 1946; *The Town,* Harmony Raine, 1950; *The Trees,* Alfred A. Knopf, 1940. Novels that detail the settlement of the Ohio River Valley, from girdling the hardwood monarchs to dispel the gloom and open corn patches, to the advent of urbanization and river commerce, to an old woman's nostalgia for the forest she despised so in her youth. Vivid images of how a continent could be transformed so fast.

RIFKIN, JEREMY, *Time Wars: The Primary Conflict in Human History,* Henry Holt, 1987. Rifkin is a philosopher who doesn't mess around. His books, of which this is the most recent, make radical inquiries into the tacit value statements of technological innovations (the nanosecond, in this case) and challenge them forthrightly. "Two futures await us," writes Rifkin, "each accompanied by its own temporal mandate. The will to power, the will to empathy. The choice is ours."

RODALE PRESS, EDITORS OF, *Earth Care for a New Decade,* Rodale Press, 1990. A large current collection of predominantly good news about what citizen activists can do to remedy environmental problems. Compiled of stories from the environmental press and other sources, the book devotes just enough space to statements of problems—acid rain, solid waste, whaling, and the like—and then generously illustrates what anyone can do to help solve them. A very positive contribution.

ROSALYN, WILL, et al., *Shopping for a Better World: A Quick and Easy Guide to Socially Responsible Supermarket Shopping,* Council on Economic Priorities, 1989. A supermarket guide

227

that ranks generally available consumer products by a number of ecological and ethical criteria.

SALE, KIRKPATRICK, *Dwellers in the Land: The Bioregional Vision,* Sierra Club Books, 1985. A suave, literate argument in favor of bioregionalism. A definitive explanation of the idea of bioregionalism by a writer and thinker who has participated in the formation of the movement, having also contributed earlier in fine works on radicalism and scale.

SAMPSON, NEIL and DWIGHT HAIR, eds., *Natural Resources for the 21st Century,* Island Press, 1989. A complete and up-to-date status report of the state of America's renewable natural resources. The book contains information gathered from America's leading resources experts on such topics as population and economic trends, climate and atmospheric trends, croplands and soil sustainability, water quantity and quality, wetlands, forestlands, and wildlife. It assesses where we stand now, forecasts what pressures will affect long-term use of our resources, and includes information on relevant legislation, as well as suggestions for the future.

SAVORY, ALLAN/CENTER FOR HOLISTIC RESOURCE MANAGEMENT, *Holistic Resource Management,* Island Press, 1988; Bingham, Sam, *The Holistic Resource Management Workbook,* Island Press, 1990. *Holistic Resource Management* is an encompassing planning model that treats people and their environment as a whole. Addressed first to ranchers, farmers, and policymakers who have primary responsibility for the health of the land, the book discusses the scientific and management principles of the model, followed by detailed descriptions of each tool and guideline. Savory's controversial, innovative, field-tested, and ecologically sound management principles (it is claimed) can be applied to a wide range of critical ecosystem problems. The companion piece, *The Holistic Resource Management Workbook,* provides the practical instruction in financial, biological, and land planning necessary to apply the holistic management model.

228

SAX, JOSEPH L., *Mountains Without Handrails: Reflections on the National Parks,* The University of Michigan Press, 1980. A lucid, historically and psychologically literate consideration of the kind of recreational activities—intensely experiencing nature or visiting a resort with scenery—that should be permitted and encouraged in U.S. national parks.

SCHAEF, ANNE WILSON, *When Society Becomes an Addict,* Harper & Row, 1987. This work examines addiction in the context of the overall societal matrix, instead of just the individual or family. Schaef finds society to have "all the characteristics and exhibit all the processes of the individual alcoholic or addict," such as the kind of pathological denial that leads to destruction of self and planet. Her analysis of how recovery is possible is a radical challenge to the patriarchal world view.

SCHELL, JONATHAN, *The Fate of the Earth,* Avon Books, 1982. About the season that would be the last—nuclear winter. A watershed book in public comprehension of just how bad nuclear war would be. Readable and frightening science. Schell's book also served to unite the peace and environmental movements behind the common cause of survival for all life forms on the planet.

SCHWEITZER, ALBERT, *Out of My Life and Thought,* Henry Holt, 1933. Selfless individual service in the twentieth century did not begin and end with Albert Schweitzer, but he was a symbol of it. A prodigiously talented organist and musicologist, as well as a theologian, Schweitzer wrote, "There came to me as I awoke the thought that I must not accept this happiness as a matter of course, but must give something in return for it." He went to medical school, went to a clinic in equatorial Africa, and struggled to arrive at a precept of the philosophy of civilization, and *reverence for life,* "the idea in which affirmation of the world and ethics are contained side by side," dawned on him like a revelation.

SEARS, PAUL B., *Deserts on the March,* 1935; Island Press, 1988. Dust-bowl-vintage writing on the wall which got read but not, finally, taken to heart. An eloquently and forcefully written history of the ecological consequences, throughout history, of mankind's habitual agricultural practices. The book portrays interactions—chains of cause and effect—and inexorable consequences and describes the squandering of the planet's legacy of fertile soils, a waste that continues unabated.

SEED, JOHN, JOANNA MACY, and ARNE NAESS, *Thinking Like a Mountain: Towards a Council of All Beings,* New Society, 1988. A sourcebook and liturgy for participatory workshops in deep ecology. Concepts and practices conducive to risking one's narrow human identity and denial in favor of opening into a sense of kinship with all imperiled life. Not for sissies, thinking like a mountain requires experiencing grief and despair on the way to hope.

SHEPARD, PAUL, *Nature and Madness,* Sierra Club Books, 1982; *The Tender Carnivore and the Sacred Game,* Charles Scribner's Sons (out of print), 1973; *Thinking Animals,* Viking Press (out of print), 1978. Shepard has been engaged in a trenchant and visionary inquiry into the historic and paleohistoric causes of our estrangement from nature for a few decades now. His books are rabid and insightful.

SHI, DAVID E., *In Search of the Simple Life: American Voices, Past and Present,* Gibbs Smith, 1986. A history of the numerous movements and communities advocating material simplicity throughout American history. This is a crosscurrent that has been flowing steady and deep. Maybe now's the time for it to become the mainstream.

SIMON, DAVID J., ed./NATIONAL PARKS AND CONSERVATION ASSOCIATION, *Our Common Lands: Defending the National Parks,* Island Press, 1988. These pages, writes Joseph L. Sax in his foreword, "constitute a virtual recipe for the protection of our parks. It's time to get cooking." The book inventories the legal means at the disposal of citizens, organizations, and park bureaucrats to preserve the environmental quality of the parks themselves.

SINGER, PETER, *Animal Liberation,* Avon Books, 1977. A first, and basic, work on the critical topic of the rights of other life forms. The book focuses largely on the cruelties of factory farming.

SLOANE, ERIC, *Almanac and Weather Forecaster,* Little, Brown (out of print), 1955. An amiable and illuminating book, attractively illustrated, which, if consulted regularly, could cultivate an accurate weather eye in any city slicker.

SNYDER, GARY, *Earth House Hold,* New Directions, 1969; *Turtle Island,* New Directions, 1974; *The Old Ways: Six Essays,* City Lights Books, 1977; *Axe Handles: Poems,* North Point Press, 1983; *Regarding Wave,* New Directions, 1970; *The Real Work: Interviews and Talks,* New Directions, 1980. Learned and spontaneous, intonable testimony from a scholar-poet/man of letters dedicated to living on the land long enough to know it well. Poetry and excellent prose that draw on many sources, including the Native American ways of the Shasta region as well as Snyder's direct experience.

SOULÉ, MICHAEL, ed., *Conservation Biology: The Science of Scarcity and Diversity,* Sinauer Associates, 1986. A compendium of mostly scientific articles that detail various aspects of the behavior of ecosystems under fire. The book reports what is becoming known about the consequences—right down to the level of the gene—of habitat disturbance on biodiversity, which is the argument for why to preserve. It relates the attempt to learn exactly how much must be preserved, along with a few preliminary suggestions for managing the preserves. Definitely not casual reading.

SOULÉ, MICHAEL E. and KATHRYN A. KOHM, eds./SOCIETY FOR CONSERVATION BIOLOGY, *Research Priorities for Conservation Biology,* Island Press, 1989. Conservation biology is a discipline with an urgent time frame. Hence the Society for Conservation Biology, in cooperation with the National Science Foundation, organized a conference of the world's leading biologists to create an agenda for research. Focusing on the need to reverse man-made changes in the atmosphere, stratosphere, and oceans, as well as on the need for conservation, *Research Priorities for Conservation Biology* proposes an immediate research agenda designed to further our understanding of the basic mechanisms that fuel and maintain biotic diversity and to increase the effectiveness of the preservation efforts.

STARHAWK, *Truth or Dare,* Harper & Row, 1987. Starhawk's analysis of power—drawn from her deep working knowledge of paganism, social activism, group dynamics, and psychology—gives readers a model of self and community based on the immanent value of all life. She suggests individual and group activities to help build nonhierarchical relationships and institutions.

STEGNER, WALLACE, *Beyond the Hundredth Meridian: John Wesley Powell and the Second Opening of the West,* University of Nebraska Press, 1982; *The Sound of Mountain Water,* University of Nebraska Press, 1985; *Mormon Country,* University of Nebraska Press, 1981. Arguably one of this century's greatest writers, Stegner has influenced the course of American writing and evolution of an environmental ethic. His works, primarily about the western United States, get at the soul of the frontier— the vast beauty of the land itself, and the determination of the people who settled it and continue to live there—and speak elegantly for the absolute necessity for conservation.

STEWART, GEORGE R., *Earth Abides,* Archive Press, 1974. A novel describing what might happen in human life if 99.9 percent of the humans succumbed to a mysterious plague, leaving a remnant population and decaying urban infrastructure. Culture reformation is what happens last.

TETRAULT, JEANNE and SHERRY THOMAS, *Country Women: A Handbook for the New Farmer,* Anchor Books (out of print), 1976. An awful lot of good advice to female back-to-the-landers. Clearly written and illustrated, *Country Women* emanated from an important magazine of the same name, which had an intelligently feminist slant.

THEOBALD, ROBERT, *The Rapids of Change: Social Entrepreneurship in Turbulent Times,* Knowledge Systems, 1987. A sagacious, distinctive description and analysis of the discontinuities of our time, rendered without blame; and seemingly simple, sophisticated ideas about new patterns of cooperation and communication that individuals and groups might consider in tackling the responsibility to usher in what Theobald, a wise man well worth knowing, dubs "the Compassionate Era."

THOMAS, WILLIAM L., JR., ed., *Man's Role in Changing the Face of the Earth, Volumes 1 and 2,* 1956; University of Chicago Press, 1971. The proceedings of an interdisciplinary conference held thirty years ago at which geographers, anthropologists, historians, archaeologists, even theologians, ecologists, botanists, and geologists addressed, as specialists, and discussed as peers the facts and implications of such human impacts as deforestation, irrigation and canal building, landscape modification by fire, disposal of waste, growth in human population, and energy use. Although global warming and ozone depletion lay beyond the conferees' horizon, their papers and symposia constitute an enduring, authoritative resource on the topic.

VAN DER POST, LAURENS, *Testament to the Bushmen,* Penguin, 1984. Van Der Post recounts his own journey to the Kalahari Desert in search of the Bushmen, who, there's reason to think, really are the first people of the Earth. Van Der Post manages to portray their relationship with the (harsh) natural world in an accessible way. Evocative, spellbinding, an entry point to hunter-gatherer's mythology, psyche, and soul.

VITTACHI, ANURADHA, *Earth Conference One: Sharing a Vision for Our Planet,* New Science Library/Shambhala, 1989. A thoughtful, subjective, and often profound account of an extraordinary forum of parliamentarians, scientists, and spiritual leaders held at Oxford in April 1988. Grappling ecumenically with planetary problems like war, hunger, and ecological crisis, this diverse gathering distilled fundamental causes—matters of heart and soul—and issued a declaration "FOR GLOBAL SURVIVAL": convergent wisdom that is so essentially fair, sufficient, and responsible that that writing job is done for now. All that remains is learning and living their vow. Highly recommended.

WACHTEL, PAUL, *The Poverty of Affluence: A Psychological Portrait of the American Way of Life,* 1983; New Society, 1988. "My hope in this book is," says author Wachtel, "to point us toward a clearer understanding of the real sources of human welfare. The growth that will improve our lives at this point is a growth in generosity and equity, two nonpolluting commodities of which we truly can never have enough." An observant, perceptive, intelligent, and principled book.

WALKER, ALICE, *Living by the Word: Selected Writings,* Harcourt Brace Jovanovich, 1988. A collection of essays, speeches, and journal entries by an eloquent, eminent writer whose passion for justice and liberation has grown outward from her revolutionary perception of the oppression of blacks and women to encompass an equally right, and deeply radical, understanding of the oppression of other living creatures, and the planet. She doesn't argue, she *knows,* and her prose has uplift and certainty polished by loving the universe whole.

WALLACE, DAVID RAINS, *The Dark Range: A Naturalist's Night Notebook,* Sierra Club Books, 1978; *Bulow Hammock: Mind in a Forest,* Sierra Club Books, 1988; *The Klamath Knot: Explorations of Myth and Evolution,* Sierra Club Books, 1983. In all his work, Wallace concerns himself with the interpenetration of landscape and mind. A great writer with a wonderful mind, Wallace depicts distinctly different places in each of his books.

WEIR, DAVID and MARK SHAPIRO, *Circle of Poison: Pesticides and People in a Hungry World,* Institute for Food and Development Policy, 1981. Early in this book, which addresses the callous, outrageous practice of our exploiting Third World markets for agricultural chemicals, we learn that *"at least 25 percent of U.S. pesticide exports are products that are banned, heavily restricted, or have never been registered for use here."* What goes around comes around, however, and we eat a lot of these poisons on imported fruits and vegetables. Concise, carefully researched, highly readable, this book expands awareness of a "detail" of the multinationalization and chemicalization of agriculture.

WESSEL, JAMES with MORT HANTMAN, *Trading the Future: The Concentration of Economic Power in Our Food System,* Institute for Food and Development Policy, 1983. A study of a market that isn't a market. Grain-trading oligopolies hiding behind the myth of the free market have finally transformed agriculture from practice for meeting basic needs, providing livelihood close to the soil, and directly dealing with subsistence into a (not infinitely) manipulable source of raw material for international economic game playing. One version of the driving force behind the catastrophic displacement of the small farmer from the countryside.

WILSON, E. O., *Biophilia: The Human Bond with Other Species,* Harvard University Press, 1984. Clearing the Amazonian rain forest for economic development "is like burning a renaissance painting to cook dinner," writes E. O. Wilson in *Biophilia.* "To explore and affiliate with life is a deep and complicated process in mental development," says Wilson. As evidenced in Wilson's writing, the process can generate a beauty of its own, an artifact such as this book, a book whose eloquence and fascination is entirely rooted in observations of the living, interacting world, especially the world of the tropics.

WORSTER, DONALD, *Nature's Economy: A History of Ecological Ideas,* Cambridge University Press, 1985. The intellectual-ecological history of the ideas of ecology. A fascinating, crisply written account of the influences playing on, and resultant colorations of, different major contributors to ecology—from Gilbert White and H. D. Thoreau to Darwin, Lyell, and Linnaeus. The book provides a cogent, authoritative basis for considering the question, what will be the role of the scientist, particularly the ecologist, in drawing designs for a revolutionary future?

ZWINGER, ANN, *A Desert Country Near the Sea: A Natural History of the Cape Region of Baja California,* Harper & Row, 1983; *Run River, Run: A Naturalist's Journey Down One of the*

235

Great Rivers of the American West, The University of Arizona Press, 1984. A prize-winning naturalist's personal guide to this ostensibly stark land. Acutely observed, extensively researched, detailed scientific observation. Zwinger possesses a rare ability to bring her subjects vividly alive for the reader and exercises it also in *Run River, Run,* the tale of her journey from the headwaters of the Green River to its confluence with the Colorado.

Reviewers' Biographies

CAROL J. ADAMS is the author of *The Sexual Politics of Meat: A Feminist-Vegetarian Critical Theory* (Continuum, 1990).

MARGOT ADLER is author of *Drawing Down the Moon* (Beacon Press, 1979). She's a reporter for National Public Radio in New York City and lectures regularly on women's spirituality.

J. BALDWIN, a student/colleague of R. Buckminster Fuller, has served as alternative technology editor for numerous editions of the *Whole Earth Catalog* and is currently doing the same for the *Whole Earth Review*.

DAVE BARRONS is a meteorologist specializing in climatology for the NBC station in Traverse City, Michigan.

MARTY BENDER was a research associate at The Land Institute for five years. He's currently a Ph.D. candidate at the University of Kentucky, Lexington, in plant ecology.

PETER BERG is one of bioregionalism's leading (and certainly most eloquent) exponents. He is director of the Planet Drum Foundation.

LINDA-RUTH BERGER is a New Hampshire poet who's been published in *American Voice, The Women's Review of Books,* and *The Sun.* She's currently working on a series of poems dedicated to Rachel Carson that she's writing from Carson's home in West South Port, Maine.

PETER BORRELLI is the editor of *The Amicus Journal,* a publication of the Natural Resources Defense Council.

DAVID R. BROWER, founder and current chairman of Earth Island Institute, has been involved in conservation battles for fifty years as Sierra Club executive director and as founder of Friends of the Earth. *For Earth's*

Sake, the first volume of his two-volume autobiography, was recently published by Peregrine Smith Books.

KATY BUTLER is a reporter for *The San Francisco Chronicle* and a longtime student of Zen Buddhism.

ERNEST CALLENBACH edits natural history guides and film books for the University of California Press. He's the author of the environmental classic *Ecotopia,* and later *Ecotopia Emerging* (see review of both books in "Spirit"), and coauthor with his wife, Christine Leefeldt, of *Humphrey the Wayward Whale* (Heyday Books, 1986).

J. BAIRD CALLICOTT is the author of *In Defense of the Land Ethic* (State University of New York Press, 1989) and editor of *Companion to a Sand County Almanac* (University of Wisconsin Press, 1987). He and Susan Flader are editing a collection of Aldo Leopold's previously unpublished papers, to be called *The Ecological Conscience and Other Essays by Aldo Leopold.*

KATHLEEN COURRIER is publications director at the World Resources Institute (see *World Resources* review in "Earth"). She's a contributor to *The Greenhouse Trap* (Beacon Press, 1990).

BARBARA DEAN is the author of *Wellspring: A Story from the Deep Country* and is executive editor of Island Press. She is also series editor of the Sierra Club Nature and Natural Philosophy Library.

JIM DODGE is the author of *Fup* (City Miner Books, 1983) and *Not Fade Away* (The Atlantic Monthly Press, 1987) as well as poetry chapbooks, columns, and short fiction. He teaches at Humboldt State University in Arcata, California, and has been a consultant in environmental restoration.

CAROL DORF's poetry and short fiction have appeared in *Contact II, Feminist Studies,* and *Heresies.* She's a former editor of *Five Fingers Review* and teaches as a California Poet in the Schools.

YAAKOV GARB studies the role of emotion in the construction of scientific knowledge. He teaches a course in environmental philosophy at the University of California, Berkeley.

LOIS MARIE GIBBS is a founder of Citizens Clearinghouse for Hazardous Waste. Her book, *Love Canal: My Story,* is reviewed in the "Water" section.

HAROLD GILLIAM is author of a dozen books on regional and environmental topics and writes a weekly environmental feature for *The San Francisco Chronicle.*

R. EDWARD GRUMBINE directs the Sierra Institute wilderness field studies program through University of California Extension at Santa Cruz.

He is also studying ecosystem management for parks and forests in the United States.

JIM HARDING directed the energy program at Friends of the Earth for thirteen years. He's currently a consultant in the energy and environment field.

JOEL W. HEDGPETH is a marine biologist, environmental activist, and poet who knew, and was greatly influenced by, the late geographer Dr. Carl Sauer. In addition to editing several textbooks on marine biology, he ran an independent marine lab for the University of the Pacific in California.

HAZEL HENDERSON is a futurist, planetary consultant on sustainability, and author (*Creating Alternative Futures*, Berkley Windhover, 1978; *The Politics of the Solar Age*, Knowledge Systems, 1981) whose long association with the late E. F. Schumacher began in 1974.

PHIL HOLLIDAY is an independent scholar-historian currently at work on a study of cycles of civilization.

DANA JACKSON is a director of education and publications for The Land Institute, which is devoted to sustainable agriculture, and edits its thrice-yearly *Land Report*.

JOE KANE is the author of *Running the Amazon* (Alfred A. Knopf, 1989), a first-person account of the first successful attempt to navigate the Amazon, by kayak, from source to mouth.

THOMAS J. LYON is a professor of English at Utah State University and editor of *This Incomperable Lande: A Book of American Nature Writing* (see Nature and Nature Writing review in "Spirit").

TIM MCNULTY is a poet (*As a Heron Unsettles a Shallow Pool*, Exiled-in-America Press, 1988), nature writer (*Washington's Wild Rivers*, Mountaineers Books, 1990), and conservationist living in the Pacific Northwest.

ROSEMARY MENNINGER ran a statewide community gardening program in California during Jerry Brown's gubernatorial administration. She now teaches art to disadvantaged teenagers at an alternative school in Topeka, Kansas.

MILTON MOSKOWITZ is a freelance business writer, author (*Global Marketplace*, Macmillan, 1988), and senior editor of *Business and Society Review*.

RICHARD NELSON is an anthropologist-become-author (*Make Prayers to the Raven*, University of Chicago Press, 1983; *The Island Within*, North Point Press, 1986) who lives near Sitka, Alaska.

ELLIOTT A. NORSE, the chief scientist for the Center for Marine Conserva-

239

tion, is author of *Conserving Biological Diversity in Our National Forests* (The Wilderness Society, 1986) and *Ancient Forests of the Pacific Northwest* (Island Press, 1990).

SEAN O'GRADY teaches English at the University of California, Davis.

TIM PALMER is the author of five books, including *Endangered Rivers and the Conservation Movement* (University of California Press, 1986) and *The Sierra Nevada* (Island Press, 1988).

DOUG PEACOCK is an author, explorer, and grizzly bear cinematographer who is rumored to have been the template for Abbey's character Hayduke.

JUDITH PLANT edited *Healing the Wounds: The Promise of Ecofeminism* (see "Further Reading") and lives in an intentional community near Lilloet, British Columbia.

PATRICIA POORE is editor and publisher of *Garbage: The Practical Journal for the Environment.*

BILL PRESCOTT is director of public information for the Climate Protection Institute, which is a project of the Earth Island Institute.

BARBARA K. RODES is research librarian for the West Wildlife Fund and The Conservation Foundation.

BARBARA SHAILOR is special projects director of the International Association of Machinists and Aerospace Workers.

DAVID SHERIDAN coauthored *A Season of Spoils* (Pantheon Books, 1984), which documents the environmental costs of the Reagan administration. He is also author of *Desertification in the United States* (Council on Environmental Quality, 1981).

DONALD SNOW edits *Northern Lights,* a quarterly publication of the Northern Lights Research and Education Institute, which analyzes issues concerning the future of the northern Rockies.

NANCY JACK TODD is coauthor of *Bioshelters, Ocean Arks, City Farming* (see review in "Water"). She edits *Annals of Earth,* a philosophical and practical journal on sustainability.

JOANNA D. UNDERWOOD is founder and president of INFORM, a national organization which identifies and reports on practical actions for the conservation of natural resources. She has lectured widely on issues of corporate environmental performance.

SIM VAN DER RYN is an architect with Sim Van Der Ryn and Associates; president of the Farralones Institute, a northern Californian center for research and education in ecologically appropriate technologies; and professor of architecture at the University of California at Berkeley. He was State Architect of California during the Brown administration. His book *Toilet Papers* is being reissued by North Atlantic Books.

CAROL VAN STRUM is a writer, author (*A Bitter Fog*, Sierra Club Books, 1983; *The Politics of Penta*, Greenpeace, 1989), and troublemaker.

DAVID RAINS WALLACE is a recipient of the prestigious John Burroughs Medal, awarded for excellence in natural history writing. His books include *The Klamath Knot* (Sierra Club Books, 1983) and *Bulow Hammock* (Sierra Club Books, 1988).

PETER WARSHALL is a biologist and ethologist who has been a land use editor for numerous editions of the *Whole Earth Catalog*.

HENRY A. WAXMAN (D-CA) is chairman of the Subcommittee on Health and Environment, U.S. House of Representatives.

MALCOLM WELLS is an architect on Cape Cod who specializes in earth-sheltered buildings.

BRIAN WILSON is a founder of the Veteran's Peace Action Team and the Institute for the Practice of Nonviolence. He lost his legs blockading arms shipments at the Concord Naval Weapons Base in California.

SETH ZUCKERMAN is a freelance writer from northern California on energy and environmental issues and coauthor with Amory and Hunter Lovins of *Energy Unbound* (Sierra Club Books, 1986).

Index

243

INDEX

Habitat destruction, 4
Haines, John, 216
Hair, Dwight, 228
Halons, 67
Halpern, Daniel, 170–71, 172
Hanemann, W. Michael, 44–45
Hansen, James, 77
Hantman, Mort, 235
Harding, Jim, 111, 239
Harper, Stephen F., 220
Hay, John, 170, 216
Hayduke Lives (Abbey), 56
Hayes, Denis, 111–12
Hays, Samuel P., 216–17
Hazardous chemicals, 81–84
Head, Suzanne, 49–51
*Headed Upstream: Interviews with
    Iconoclasts* (Loeffler), 219
Healing the Wounds (Plant), 226
Hedgpeth, Joel W., 16–18, 239
Heinzman, Robert, 49–51
Henderson, Hazel, 186–87, 217, 239
Hildyard, Nicholas, 216
Historical geography, 18
History, 15–16
Hoagland, Edward, 170, 171
Holdren, John P., 214
Holistic Resource Management (Savory),
    228
*Holistic Resource Management
    Workbook, The* (Savory), 228
Holliday, Phil, 32–33, 177–78, 239
Home heating, 97–98
Home Planet, The (Kelley), 71–73
*Home to the Wilderness: A Personal
    Journey* (Carrighar), 212
Hopi culture, 158–60, 173
Hopi Kachinas (Wright), 158–59
House of Life (Brooks), 146–47
*How Can I Help? Stories and Reflections
    on Service* (Ram Dass and Gorman),
    200–203
Huls, Jon, 151–53
Hung-jen, 106, 107
Hunt, Constance Elizabeth, 217
Huntley, Brian J., 45
Hur, Robin, 40
Huser, Verne, 217
Huxley, Aldous, 103
Hyams, Edward, 28–30
Hynes, H. Patricia, 217

I and Thou (Buber et al.), 211
Ignorance, 162
Iltis, Hugh, 44
In Defense of Nature (Hay), 216
Indian Country (Matthiessen), 221
Indian Tales (De Angulo), 213

Industrialization, 101–105, 129,
    176–77
*In Search of the Simple Life: American
    Voices, Past and Present* (Shi), 230
Integration, 163
Intensive gardening, 37
International Green Front Report (Friends
    of the Trees), 107, 108, 109
*In the Rainforest: Report from a Strange,
    Beautiful, Imperiled World* (Caufield),
    49
Invention, 92
Irrigation, 126, 135–36
*Ishi in Two Worlds: A Biography of the
    Last Wild Indian in North America*
    (Kroeber), 218
Island Within, The (Nelson), 224

Jackson, Dana, 30–31, 239
Jackson, Wes, 32, 33, 34, 35, 94
Jacobsen, C. B., 135
Jahner, Elaine, 173
Janzen, Daniel, 15, 44
Jenkins, Bob, 44
John Muir: To Yosemite and Beyond
    (Muir), 167
*John of the Mountains: The Unpublished
    Journals of John Muir* (Muir), 167
Johnson, Warren, 217
Journals of Lewis and Clark, The
    (DeVoto), 213

Kachina dolls, 158–59
Kane, Joe, 49–50, 239
Kazis, Richard, 84–86
Kelley, Kevin W., 71–73
Kenya, 52–53
*Killing Rain: The Global Threat of Acid
    Precipitation, A* (Pawlick), 225
King, Martin Luther, Jr., 162
*Klamath Knot: Explorations of Myth and
    Evolution, The* (Wallace), 234
Kohm, Kathryn A., 231–32
Kohr, Leopold, 218
Kroeber, Theodora, 218
Krutch, Joseph Wood, 218
Kuchenberg, Tom, 136, 137

Lame Deer, 205
Land and Life (Sauer), 16, 18
Land-Saving Action (Brenneman and
    Bates), 211
*Landscape, History and the Pueblo
    Imagination* (Silko), 173
Lao Tsu, 207
Lappé, Frances Moore, 38–39
Last Stand of the Red Spruce, The (Mello),
    221–22

247

Acknowledgments

Thanks to Island Press, and its president, Chuck Savitt, for the invitation to do this project and to Barbara Dean, Island Press executive editor, for her unfailing nurture and keen editorial judgment; thanks to Jeanne Carstensen for bringing skill, verve, high energy, and good politics to the large task of editing the book reviews and "Further Reading."

Thanks as well to Sally Van Vleck of the Neahtawanta Research and Education Center for giving me permission to let my garden go to pot in order to devote the summer of '89 (and fall and winter) to this worthwhile endeavor; and Bob Russell, also of the Neahtawanta Center, for all manner of help, from reviewing manuscripts of the essays, to recommending books for review and bibliography, to generously lending background materials.

Thanks to Sue Flerlage of the flying fingers, who typed everything and provided clerical assistance, all with a good heart and steady mind from start to finish. Amazing!

Thanks to Denis Hayes, chief executive officer of Earth Day 1990, for consenting to be interviewed and for devoting a lot of his scarcest resource—time—to making that interview excellent.

Thanks to the Leland Library for allowing me to haul off entire shelves full of books and for being remarkably forgiving when I kept them overdue; Bill Rastetter for letting his home office infrastructure be used to process these words, and both Bill and Andy Rastetter for coming to my rescue at interview time with a colorful and effective

253

My First Sony tape recorder; Dave Barrons, meteorologist at the NBC station in Traverse City, for weather education; Captain Tom Kelly of the Inland Seas Education Association for conning the water chapter; Rob Chapman of Petrostar Energy for clarifying the geologic eras; and Jeff Forrest of the Computer Haus in Traverse City for sacrificing his own personal Mac to be used in the last hours of preparing this manuscript.

—STEPHANIE MILLS

I'd like to express my thanks to the following people—who each helped greatly in the execution of this complicated project: Point Foundation/*Whole Earth Review* for use of their incomparable library and the support of the staff, who had to put up with my moonlighting, often fielding FAXes and phone calls for me; Yaakov Garb for ongoing invaluable discussion about environmental ethics (someday I'll take your course!); Pat Murphy and Richard Kadrey for compiling a list of eco-science fiction books; Seth Zuckerman for taking the time to discuss energy issues with me; J. Baldwin for the many benefits of our first collaboration (and for giving me more good leads and commonsense advice about this project); Peter Warshall for wise counsel about current thinking in environmental conservation in developing countries; Richard Nilsen for many spontaneous and illuminating conversations about books, ideas, and resources pertinent to this book; Keith Jordan for his generous help designing hypercard stacks and otherwise keeping me computer-sane; the folks at Heyday Books for their persistence in helping me acquire a copy of Carl Sauer's *Man in Nature;* and Richard Schauffler, my roommate, for tolerating a very busy "home" phone.

I'd also like to thank all the contributors to this book. Each one agreed generously to work on a tight deadline mostly for love of the subject. That there are so many knowledgeable people willing to write simply for nature's sake is encouraging, to say the least.

—JEANNE CARSTENSEN

About the Editors

STEPHANIE MILLS is a writer and editor whose career began in 1970 at the helm of *EarthTimes,* a muckraking tabloid. Mills also served as editor of *Not Man Apart, CoEvolution Quarterly,* and *California Tomorrow.*

Her first book, *Whatever Happened to Ecology?,* was published by Sierra Club Books in 1989. Her work appears in the *Whole Earth Review* and *Utne Reader* from time to time. She lives near Maple City, Michigan.

JEANNE CARSTENSEN is a writer and editor. With Stewart Brand, she coedited the weekly *Chronicle Whole Earth Catalog* column appearing in the *San Francisco Chronicle.* Carstensen guest-edited the winter 1989 and winter 1987 issues of the *Whole Earth Review.* She was managing editor of *The Essential Whole Earth Catalog,* published by Doubleday in 1986. Before that she was codirector of the Planet Drum Foundation.

From her home base in San Francisco, Carstensen has traveled extensively in Nepal, Mexico, and Central America.

Also Available from Island Press

Ancient Forests of the Pacific Northwest
By Elliott A. Norse

Balancing on the Brink of Extinction: The Endangered Species Act and Lessons for the Future
Edited by Kathryn Kohm

Better Trout Habitat: A Guide to Stream Restoration and Management
By Christopher J. Hunter

The Challenge of Global Warming
Edited by Dean Edwin Abrahamson

Coastal Alert: Ecosystems, Energy, and Offshore Oil Drilling
By Dwight Holing

The Complete Guide to Environmental Careers
The CEIP Fund

Creating Successful Communities: A Guidebook to Growth Management Strategies
By Michael A. Mantell, Stephen F. Harper, and Luther Propst

Crossroads: Environmental Priorities for the Future
Edited by Peter Borrelli

Economics of Protected Areas
By John A. Dixon and Paul B. Sherman

Environmental Restoration: Science and Strategies for Restoring the Earth
Edited by John J. Berger

Fighting Toxics: A Manual for Protecting Your Family, Community, and Workplace
By Gary Cohen and John O'Connor

Hazardous Waste from Small Quantity Generators
By Seymour I. Schwartz and Wendy B. Pratt

Last Stand of the Red Spruce
By Robert A. Mello

Natural Resources for the 21st Century
Edited by R. Neil Sampson and Dwight Hair

The New York Environment Book
By Eric A. Goldstein and Mark A. Izeman

257

Overtapped Oasis: Reform or Revolution for Western Water
By Marc Reisner and Sarah Bates

Plastics: America's Packaging Dilemma
By Nancy Wolf and Ellen Feldman

The Poisoned Well: New Strategies for Groundwater Protection
Edited by Eric Jorgensen

Race to Save the Tropics: Ecology and Economics for a Sustainable Future
Edited by Robert Goodland

Recycling and Incineration: Evaluating the Choices
Edited by Richard A. Denison and John Ruston

Research Priorities for Conservation Biology
Edited by Michael E. Soulé and Kathryn Kohm

Resource Guide for Creating Successful Communities
By Michael A. Mantell, Stephen F. Harper, and Luther Propst

The Rising Tide: Global Warming and World Sea Levels
By Lynne T. Edgerton

Rivers at Risk: The Concerned Citizen's Guide to Hydropower
By John D. Echeverria, Pope Barrow, and Richard Roos-Collins

Rush to Burn: Solving America's Garbage Crisis?
From *Newsday*

Shading Our Cities: A Resource Guide for Urban and Community Forests
Edited by Gary Moll and Sara Ebenreck

War on Waste: Can America Win Its Battle with Garbage?
By Louis Blumberg and Robert Gottlieb

Western Water Made Simple
From *High Country News*

Wetland Creation and Restoration: The Status of the Science
Edited by Mary E. Kentula and Jon A. Kusler

Wildlife of the Florida Keys: A Natural History
By James D. Lazell, Jr.

For a complete catalog of Island Press publications, please write:
    Island Press
    Box 7
    Covelo, CA   95428
    (1-800-828-1302)